Springer Theses

Recognizing Outstanding Ph.D. Research

Aims and Scope

The series "Springer Theses" brings together a selection of the very best Ph.D. theses from around the world and across the physical sciences. Nominated and endorsed by two recognized specialists, each published volume has been selected for its scientific excellence and the high impact of its contents for the pertinent field of research. For greater accessibility to non-specialists, the published versions include an extended introduction, as well as a foreword by the student's supervisor explaining the special relevance of the work for the field. As a whole, the series will provide a valuable resource both for newcomers to the research fields described, and for other scientists seeking detailed background information on special questions. Finally, it provides an accredited documentation of the valuable contributions made by today's younger generation of scientists.

Theses are accepted into the series by invited nomination only and must fulfill all of the following criteria

- They must be written in good English.
- The topic should fall within the confines of Chemistry, Physics, Earth Sciences, Engineering and related interdisciplinary fields such as Materials, Nanoscience, Chemical Engineering, Complex Systems and Biophysics.
- The work reported in the thesis must represent a significant scientific advance.
- If the thesis includes previously published material, permission to reproduce this must be gained from the respective copyright holder.
- They must have been examined and passed during the 12 months prior to nomination.
- Each thesis should include a foreword by the supervisor outlining the significance of its content.
- The theses should have a clearly defined structure including an introduction accessible to scientists not expert in that particular field.

More information about this series at http://www.springer.com/series/8790

Rui Sun

An Integrated Solution Based Irregular Driving Detection

Doctoral Thesis accepted by
Imperial College London, UK

Springer

Author
Dr. Rui Sun
College of Civil Aviation
Nanjing University of Aeronautics and Astronautics
Nanjing, Jiangsu Province
China

Supervisor
Prof. Washington Yotto Ochieng
Department of Civil and Environmental Engineering
Imperial College London
London
UK

Co-Supervisors
Dr. Shaojun Feng
Department of Civil and Environmental Engineering
Imperial College London
London
UK

Dr. Wolfgang Schuster
Department of Civil and Environmental Engineering
Imperial College London
London
UK

ISSN 2190-5053 ISSN 2190-5061 (electronic)
Springer Theses
ISBN 978-3-319-44925-8 ISBN 978-3-319-44926-5 (eBook)
DOI 10.1007/978-3-319-44926-5

Library of Congress Control Number: 2016949124

Printed on acid-free paper

This Springer imprint is published by Springer Nature
The registered company is Springer International Publishing AG
The registered company address is: Gewerbestrasse 11, 6330 Cham, Switzerland

Supervisor's Foreword

Vehicle state estimation using Global Navigation Satellite Systems (GNSS) is a core element of Intelligent Transport Systems (ITS), with the potential to make a significant contribution to tackling transport-related problems including safety. Although significant advances have been made, the state-of-the-art state estimation methods struggle to support the more demanding applications, especially those that require high accuracy positioning. For example, centimetre/decimetre-level positioning accuracy is required for some safety-related applications including lane control, collision avoidance, and advanced driver assistance (ADAS) systems. This book addresses safety-related lane-level applications represented by the capability to detect lane-level irregular driving behaviour, a key cause of incidents and accidents.

The book starts with a critical review of the literature on the state of the art of irregular driving detection methods. It identifies the main limitations as (1) the lack of high accuracy positioning information and vehicle dynamic parameters; and (2) inability to extract irregular driving patterns from the current low accuracy positioning information and dynamic parameters. The book addresses these two limitations by specifying and testing a novel-integrated solution for the detection of irregular driving.

Due to the complexity of the problem, the book develops the solution into two steps. The first develops a new algorithm for high accuracy positioning based on the integration of GNSS with an Inertial Navigation System (INS), using particle filtering with precise vehicle motion models and lane information. The second step specifies a Fuzzy Inference System (FIS) for use with information from the second step to detect different types of irregular driving. The solution is tested using a combination of simulated and field data and shown to be practical and effective in achieving the desired outcomes. Overall, this book makes a significant contribution to the improvement in safety through the detection of irregular driving and is transferable to other applications that require high accuracy positioning information.

London, UK
May 2016

Prof. Washington Yotto Ochieng, FREng

Abstract

Global Navigation Satellite Systems (GNSS) are widely used in the provision of Intelligent Transport System (ITS) services. Today, metre-level positioning accuracy, which is required for many applications including route guidance, fleet management and traffic control, can be fulfilled by GNSS-based systems. Because of this level of success and potential, there is an increasing demand for GNSS to support applications with more stringent positioning requirements. These include safety-related applications that require centimetre/decimetre-level positioning accuracy, with high integrity, continuity and availability such as lane control, collision avoidance and intelligent speed assistance. Detecting lane-level irregular driving behaviour is the basic requirement for lane-level ITS applications.

Currently, some research has addressed road-level irregular driving detection; however, very little research has been done in lane-level irregular driving detection. The two major issues involved in the lane-level irregular driving identification are access to high accuracy positioning and vehicle dynamic parameters, and extraction of erratic driving behaviour from this and the lane-related information.

This thesis proposes an integrated solution for the detection of lane-level irregular driving behaviour. Access to high accuracy positioning is enabled by GPS and its integration with an Inertial Navigation System (INS) using extended Kalman filtering (EKF) and particle filtering (PF) with precise vehicle motion models and lane centre line information. Four motion models are used in this thesis: constant velocity (CV), constant acceleration (CA), constant turn rate and velocity (CTRV) and constant turn rate and acceleration (CTRA). The CV and CA models are used on straight lanes and the CTRV and CTRA models on curved lanes. Lane centre line information is extracted from defined lane coordinates in the simulation and is surveyed and stored as sets of positioning points from the motorway in the field test. The high accuracy vehicle positioning and dynamic parameters include yaw rate (*omega*) and lateral displacement (*d*) in addition to conventional navigation parameters such as position, velocity and acceleration.

The detection of irregular driving behaviour is achieved by comparing the sorting rules of a driving classification indicator from the filter estimations with what is extracted from the reference. The detected irregular driving styles are characterized by weaving, swerving, jerky driving and normal driving on straight and curved lanes, based on the Fuzzy Inference System (FIS).

The solution proposed in the thesis has been tested by simulation and validated by real field data. The simulation results show that different types of lane-level irregular driving behaviour can be correctly identified by the algorithms developed in this thesis. This is confirmed by the application of data from a field test during which the dynamics of an instrumented vehicle supplied by Imperial College London were captured in real time. The results show that the precise positioning algorithms developed can improve the accuracy of GPS positioning and that the FIS-based irregular driving detection algorithms can detect the different types of irregular driving. The evaluation of the designed integrated systems in the field test shows that a positioning accuracy of 0.5 m (95 %) source is required for lane-level irregular driving detection, with a correct detection rate of 95 % and availability of 94 % based on a 1 s output rate. This is useful for many safety-related applications including lane departure warnings and collision avoidance.

Publications Related to This Thesis

Sun, R., Ochieng, W.Y. and Feng, S. (2015). An integrated solution for lane level irregular driving detection on highways. Transportation research part C: Emerging technologies. Volume 56, July 2015, Pages 61–79. DOI:10.1016/j.trc.2015.03.036.

Sun, R., Ochieng, W., Fang, C. and Feng, S. (2015). A New Algorithm for Lane Level Irregular Driving Identification. The journal of navigation. Volume 68, Pages 1173–1194. DOI:10.1017/S0373463315000491.

Sun R., Feng, S. and Ochieng, W.Y. (2014). "An Integrated Solution Based Irregular Driving Detection," Proceedings of the 27th International Technical Meeting of The Satellite Division of the Institute of Navigation (ION GNSS+ 2014), Tampa, Florida, September 2014, pp. 548–558.

Acknowledgements

Firstly, I would like to express my sincere gratitude to my main supervisor Prof. Washington Ochieng who not only offered me the opportunity and a conducive environment to carry out this research, but also set a vivid example of a good researcher and great group leader. Without his academic support, this thesis could not have been accomplished.

I would also like to express my sincere gratitude to my other supervisors Dr. Shaojun Feng and Dr. Wolfgang Schuster, for the many useful discussions, as well as for their dedicated efforts in revising this thesis. In addition, I would like to thank the following past and current members of the Departments of Civil and Environmental Engineering, and Electrical and Electronic Engineering Department, Dr. Yanyang Zhang, Dr. Libin Fang, Dr. Xin Liu, Dr. Ke Han, Dr. Jiang Han, Dr. Fangce Guo, Dr. Jun Hu, Mr. Jianlin Luan, Dr. Haishi Ning and Dr. Yili Xia, for their support. Ms. Fionnuala Donovan and Mrs. Jackie Sime are acknowledged for providing excellent administrative support.

Finally, I would like to gratefully acknowledge the support and help of all my other colleagues at the Imperial College London.

Declaration of Contribution

I confirm that I am aware of the Imperial College London's policy on plagiarism and I certify that all the work, analysis and opinions presented in this dissertation are of my own except for:

Most of the information contained in Chaps. 2 and 3 was taken from books, research papers, conference proceedings and electronic sources, which are original of their authors and which, to the best of my knowledge, were accordingly cited when required and referenced in the Reference section of this dissertation.

In addition, the software, equipment and processes used during this thesis which were not designed by me have been appropriately acknowledged.

Copyright Declaration

Contents

List of Figures

List of Tables

Acronyms

ABS	Antilock braking system
ADAS	Advanced driver assistance systems
ATIS	Advanced traveller information systems
ATMS	Advanced transport management ystems
CA	Constant acceleration
CCA	Constant curvature and acceleration
CCD	Charge coupled device
CSAV	Constant steering angle and velocity
CTRA	Constant turn rate and acceleration
CTRV	Constant turn rate and velocity
CV	Constant velocity
DGPS	Differential GPS
DSRC	Dedicated short-range communications
ECU	Engine control unit
EKF	Extended Kalman filter
EKFCA	Extended Kalman filter constant acceleration
EKFCV	Extended Kalman filter constant velocity
FIS	Fuzzy Inference System
FNN	Fuzzy neural networks
GIS	Geographic Information System
GNSS	Global Navigation Satellite System
GPS	Global Positioning System
HMM	Hidden Markov model
ICR	Instantaneous centre of rotation
IMU	Inertial Measurement Unit
INS	Inertial Navigation System
ISA	Inertial sensor assembly
ITS	Intelligent Transport Systems
IVIS	In-Vehicle Information and Communication Systems
KF	Kalman filter

LD	Large D-indicator
LDWS	Lane departure warning system
LO	Large O-indicator
LSQ	Least squares
MAD	Moving average deviation
MD	Medium D-indicator
MEMS	Micro-electromechanical system
MO	Medium O-indicator
NHTSA	National Highway Traffic Safety Administration
NRTK	Network-based real-time kinematic
OBD	On-board diagnosis
OBU	On-board Units
OTF	On-the-fly
PDF	Probability density function
PF	Particle filter
PFCA	Particle filter and constant acceleration
PFCV	Particle filter and constant velocity
PNT	Positioning navigation and timing
PPP	Precise point positioning
RF	Radio frequency
RTCM	Radio Technical Communication for Maritime Services
RTK	Real-time kinematic
SD	Small D-indicator
SIR	Sampling importance resampling
SMC	Sequential Monte Carlo
SO	Small O-indicator
TOC	Traffic operations centres
TTA	Time to alarm
VLD	Very large D-indicator
VLO	Very large O-indicator

Chapter 1
Introduction

Abstract To date, there has been very little improvement in the ability to detect lane level irregular driving styles, mainly due to a lack of high performance positioning techniques and suitable driving pattern recognition algorithms. This chapter sets out the problems that currently exist in regard to these safety issues, and the current focus of research to address these problems. This is followed by the identification of the relevant key issues for this thesis and the formulation of its objectives. The chapter ends by outlining the structure of the rest of the thesis.

Transport is a vital element in the development of society. In recent decades, as people's living conditions have improved and their disposable income has grown, demand for vehicles has surged rapidly. In 2013, there were 35 million vehicles licensed for use on the road in Great Britain, up 40 % from 25 million in 1994 (DfT 2014). This large growth in the use of road transport has brought many benefits, including stimulating global economic growth. This has been accompanied, however, by an increased risk of traffic accidents, with enormous associated costs, both human and financial. Reducing the occurrence of accidents is, therefore, a matter of paramount concern requiring the enhancement of drivers' safety consciousness, but also the adoption of technical measures to warn and prevent drivers from making errors.

The detection of irregular driving has become a major issue in modern transport safety. Accidents generally do not occur randomly; instead, there is usually an identifiable pattern of behaviour in the period prior to an accident (Chang et al. 2008). It is necessary, therefore, to develop a system capable of the early detection of different types of irregular driving patterns within the lane in order to warn the driver in sufficient time to avoid an accident.

To date, there has been very little improvement in the ability to detect lane level irregular driving styles, mainly due to a lack of high performance positioning techniques and suitable driving pattern recognition algorithms. The remainder of this chapter sets out the problems that currently exist in regard to these safety issues, and the current focus of research to address these problems. This is followed by the

R. Sun, *An Integrated Solution Based Irregular Driving Detection*,
Springer Theses, DOI 10.1007/978-3-319-44926-5_1

identification of the relevant key issues for this thesis and the formulation of its objectives. The chapter ends by outlining the structure of the rest of the thesis.

1.1 Background

Safety is one of the main transport related societal problems in addition to congestion and pollution. In 2012, the total number of casualties in road accidents in the UK was 195,723, of which 1754 were fatal and 23,039 were serious injuries (Kilbey 2013). Abnormal driving behaviours, including sudden lane changes and erratic driving due to fatigue and drowsiness, account for more than 90 % of these accidents (Aljaafreh et al. 2012).

These driving styles, which are characterized as weaving, swerving and jerky driving, are considered as irregular driving by the U.S. National Highway Traffic Safety Administration (NHTSA) (NHTSA 2010). Early detection of irregular driving within the lane has the potential to prevent the occurrence of accidents. This capability could be provided by the exploitation of Intelligent Transport Systems (ITS) technologies to realize irregular driving identification (for evaluating driving performance and improving driving behaviour), and to enable potential benefits for lane control and collision avoidance applications. To date, research focusing on the detection of irregular driving is mainly based on two approaches, namely, the detection of the vehicle's real-time driving pattern and the monitoring of the driver's physical behaviour.

For the detection of real-time driving patterns, various sensors, such as positioning, orientation, velocity and vision, are used to detect vehicle motion information, and the collected data is subsequently analysed to find cues for irregular driving.

In regard to the research on vehicle driving pattern detection, different sensors including positioning, orientation, velocity and vision are used to capture vehicle motion information. The analysis of the collected data is used to find cues for irregular driving.

Lecce and Calabrese (2008) designed a driving information collection system based on specific sensors and GPS receivers and a pattern matching algorithm to classify driving styles. Their research was preliminary, however, and did not present any quantitative simulation and/or field test results. Chang et al. (2008) developed a vision based system which was able to learn the trajectories and longitudinal and lateral velocities of the vehicle based on Fuzzy Neural Networks (FNN) and, subsequently, to calculate the level of danger posed by the vehicle. The performance was assessed by both simulations and field tests. Issues with this system included the fact that the research only classified the risk level as safe, cautious or dangerous, and that the vision sensor's performance was shown to be affected by the weather during the field test. Imkamon et al. (2008) proposed a vision and orientation sensor based method for detecting unsafe driving behaviour. A fuzzy logic system was subsequently used to combine the measured data and classify different levels of hazardous driving. The levels of hazardous driving were only roughly classified by a self-defined risk rating, however, and the performance was not quantified clearly. Krajewski et al. (2009), meanwhile, extracted driver fatigue information from steering behaviour based on collected steering data from orientation sensors before using signal processing to capture fatigue impaired patterns. The research simply classified the driving styles as fatigue or non-fatigue and the performance of the system was highly affected by the accuracy of the collected data. Dai et al. (2010) proposed a system to detect drunk-driving manoeuvres using a mobile phone with an accelerometer and orientation sensor. The acceleration patterns from sensor readings were computed to match the typical drunk-driving patterns extracted from real driving tests. The correct detection rate was not quantified, however, although it was argued that the system performance could be improved if GPS information was integrated. Saruwatari et al. (2012) developed a vision sensor based method for detecting abnormal driving of vehicles. The abnormal vehicle motions could be extracted in the form of group behaviour by using a multi-linear relationship in space-time images. This research, however, only generally described the algorithm and did not present either a simulation or field experiment results. Table 1.1 summarizes current research based on real-time driving pattern detection approach by the equipment, algorithm and irregular driving detection performance. It indicates that most of the research to classify irregular driving types to date is preliminary and does not attempt to quantify performance in terms of the time to first detection, availability and correct detection rate of irregular driving.

Although the real-time driving pattern detection approach has shown great potential for detecting irregular driving, a technical barrier that needs to be surmounted is the fact that the performance of vision sensor based research can be affected by weather conditions. In addition, some studies employing this approach rely significantly on the readings of high grade GPS or other motion sensors, the cost of which may hinder their widespread practical use. Moreover, most of the irregular driving detection systems discussed above are still at an early stage of development, with neither field tests nor a robust algorithm to distinguish or classify different types of irregular driving styles. Their efficiency and reliability need to be further examined, therefore.

On the other hand, visual or auxiliary systems are always used to monitor the driver's physical behaviour. Visual observation is, therefore, an option for detecting driver fatigue. Eriksson and Papanikolopoulos (2001) developed a vision-based approach for diagnosing driver fatigue by monitoring a driver's eyes and issuing a warning when irregular eye closure was detected. Zhu and Ji (2004) developed a two-camera based method to predict fatigue with a probabilistic model based on the captured visual cues of drivers, such as eyelid movement, gaze movement, head movement and facial expression. Lee et al. (2006) also developed a camera based method to monitor a driver. The camera was used to capture the driver's sight line and driving path and the correlation between them was subsequently calculated. Omidyeganeh et al. (2011) proposed a scheme for driver drowsiness detection based on a fusion of eye closure and yawning detection methods. The driver's facial

Table 1.1 Summary of irregular driving detection based on real-time driving pattern detection approach

Equipment	Algorithm	Irregular driving classification results evaluation				Reference
		Time to first detection	Irregular driving type	Availability	Correct detection rate	
GPS and data acquisition device	Fuzzy logic	N/A	N/A	N/A	N/A	Lecce and Calabrese (2008)
CCD camera	Fuzzy neural networks	N/A	Safe, cautious, danger	N/A	Lane departure: 96.34 % on simulation and 89.29 % on field test collision avoidance: 100 % on simulation and 80 % on field test	Chang et al. (2008)
Vision, orientation and acceleration sensor	Fuzzy logic	N/A	Self-defined risk grade number from 1 to 3	N/A	N/A	Imkamon et al. (2008)
Orientation sensor	Signal processing	N/A	Fatigue or not fatigue	N/A	86.1 % for Fatigue detection	Krajewski et al. (2009)
Accelerometer and orientation sensor	Pattern matching	N/A	Drunk or not drunk	N/A	N/A	Dai et al. (2010)
Vision sensors	Image processing	N/A	Self-defined degree	N/A	N/A	Saruwatari et al. (2012)

appearance was captured via a camera installed in the car and the signs of driver fatigue were studied. Once the driver's state of drowsiness was determined, a warning was sent to the driver.

Apart from the above investigations, auxiliary systems coupled with the vehicle for the detection of vehicle-driver interactions is also an option for monitoring the driver. Albu et al. (2008) not only used a vision-based system to monitor eye conditions in order to detect fatigue while driving but also used an auxiliary force sensor on the accelerator pedal and collected the exerted force to monitor driver fatigue. Heitmann et al. (2001) proposed a multi-parametric approach to monitor and prevent driver fatigue based on various auxiliary sensors, including a head position sensor, an eye-gaze system, a two pupil-based system and an in-seat vibration system, for alertness monitoring. Desai and Haque (2006) designed a system to define the level of driver alertness based on the time derivative of the force exerted by the driver at the vehicle-human interface, such as pressure on the accelerator pedal. Sandberg et al. (2011) developed a physiological signals based method for abnormal driving detection. Electroencephalography was used to detect brain activity and signs of sleepiness were extracted by analysing the signal.

While monitoring of the driver's physical behaviour has emerged as a potential approach, the vision sensor installed inside the vehicle can lead to safety hazards due to driver distractions. Moreover, ambiguous road markings and bad weather can seriously affect the performance of the vision sensor. For the auxiliary system, compatibility issues can be raised since a coupled system is strictly required for system operation. Furthermore, the complexity and high cost of the auxiliary systems can make it difficult to integrate in practice. Thus, the irregular driving detection approach based on physical monitoring of the driver is difficult to apply widely.

From the discussion of the two potential approaches for irregular driving detection, it can be seen that the physical monitoring of the driver approach is difficult for real applications, while the approach based on detection of the driving pattern is potentially feasible, especially since the performance of such a system can be improved by more accurate positioning and more advanced irregular driving identification algorithms.

1.2 Aims and Objectives

From the issues discussed above, the aim of the thesis is to propose a new lane level irregular driving detection algorithm based on a real-time driving pattern detection approach. The algorithm includes two parts. The first part is a GPS/INS integration algorithm to provide lane level positioning and highly accurate estimation of dynamic parameters, while the second part is Fuzzy Inference System (FIS) based irregular driving identification algorithm to distinguish different types of irregular driving. In order to achieve these aims, the following objectives are pursued in this thesis:

- To carry out an extensive literature review to capture the state-of-the-art in the existing irregular driving monitoring algorithms.
- To identify and provide solutions to address basic underpinning issues such as system design, filter choice, vehicle motion model choice and driving pattern detection methods.
- To develop advanced EKF and PF models for GPS and INS integration with different motion models.
- To develop an advanced FIS based algorithm for the classification of driving patterns.
- To establish a Matlab and Simulink based simulation platform with an efficient data processing structure for testing irregular driving identification systems based on multi-combination models.

- To provide analysis of the performance of the novel irregular driving detection algorithms, employing both straight lane and curved lane data, and simulated/or real data where practical.

This thesis is mainly focused on vehicles moving mainly within the lanes of motorways and A-roads with sufficient GPS data. Although the thesis seeks primarily to develop algorithms for irregular driving identification within a lane, the algorithms should be easily transferable to other applications that require real-time high accuracy positioning and pattern recognition.

1.3 Structure of the Thesis

The thesis is organized into six chapters, focusing on six topics. Each of the chapters begins with an overview, followed by a number of sub-sections.

Chapter 1: Introduction
The first chapter presents the overall context of the research together with the objectives and structure of the thesis.

Chapter 2: Road Safety and Intelligent Transport Systems
This chapter reviews the state-of-the-art in road traffic accidents and safety, as well as related ITS concepts for safety applications. The role of Positioning Navigation and Timing (PNT), which includes the underpinning technologies for ITS, is introduced with GNSS technology in particular being highlighted. Overall, this chapter captures the background information both in terms of the practical relevance of the research presented in this thesis and its objectives.

Chapter 3: State-of-the-Art in Irregular Driving Detection
This chapter introduces current monitoring and detection methods for irregular driving. The definition and benefits of irregular driving detection are reviewed. This is followed by a review and consideration of the shortcomings of two categories of irregular driving identification methods: the detection of a vehicle's real-time driving pattern and the monitoring of drivers' physical behaviour. In-car positioning is also reviewed to introduce the issue of sensor integration by filters for high accuracy estimation of vehicle positioning and related parameters. Finally, driving pattern recognition algorithms are reviewed.

Chapter 4: A New System for Lane Level Irregular Driving Detection
This chapter presents the developments in this thesis. Versatile models (including a kinematic model and filter model combination) are designed, both on straight and curved lanes, and the FIS based driving pattern recognition model are specified.

Chapter 5: Testing, Analysis and Performance Validation
This chapter defines several scenarios both on straight and curved roads. Based on these scenarios, the sorting rules of driving classification indicators used to distinguish different types of irregular driving styles, are extracted. The results of

the algorithms developed are compared by simulation, before testing with real field data.

Chapter 6: Conclusions and Recommendations for Future Work
Based on the work presented in this thesis, this chapter summarises the results achieved and their impact, and provides recommendations for future work.

References

Albu, A. B., Widsten, B., Wang, T., Lan, J., & Mah, I. (2008). A computer vision-based system for real-time detection of sleep onset in fatigued drivers. *IEEE Intelligent Vehicles Symposium, 2008*, 25–30.

Aljaafreh, A., Alshabatat, N., & Najim Al-Din, M. S. (2012). Driving style recognition using fuzzy logic. *IEEE International Conference on Vehicular Electronics and Safety (ICVES), 2012*, 460–463.

Chang, T. H., Hsu, C. S., Wang, C., & Yang, L. K. (2008). Onboard measurement and warning module for irregular vehicle behaviour. *IEEE Transactions on Intelligent Transportation Systems, 9*(3), 501–513.

Dai, J., Teng, J., Bai, X., Shen, Z., & Xuan, D. (2010). Mobile phone based drunk driving detection. In *2010 4th international conference on pervasive computing technologies for healthcare (Pervasive Health)* (Vol. 1).

Desai, A. V., & Haque, M. A. (2006). Vigilance monitoring for operator safety: A simulation study on highway driving. *Journal of Safety Research, 37*(2), 139–147.

DfT. (2014). *Vehicle licensing statistics: 2013 Report.*

Eriksson, M., & Papanikolopoulos, N. P. (2001). Driver fatigue: A vision-based approach to automatic diagnosis. *Transportation Research Part C: Emerging Technologies, 9*(6), 399–413.

Heitmann, A., Cuttkuhn, R., Aguirre, A., Trutschel, U., & Moore-Ede, M. (2001). Technologies for the monitoring and prevention of driver fatigue. In *Proceedings of the fifth international driving symposium on human factors in driver assessment, training and vehicle design* (pp. 81–86).

Imkamon, T., Saensom, P., Tangamchit, P., & Pongpaibool, P. (2008). Detection of hazardous driving behavior using fuzzy logic. In *5th International conference on electrical engineering/electronics, computer, telecommunications and information technology, ECTI-CON 2008* (Vol. 2, p. 657).

Kilbey, P. (2013). *Reported road casualties in Great Britain: 2012 annual report.* Department of Transport.

Krajewski, J., Sommer, D., Trutschel, U., Edwards, D., & Golz, M. (2009). Steering wheel behavior based estimation of fatigue. In *Proceedings of the fifth international driving symposium on human factors in driver assessment, training and vehicle design* (pp. 118–124).

Lecce, V. D., & Calabrese, M. (2008). Experimental system to support real time driving pattern recognition. In *Advanced intelligent computing theories and applications with aspects of artificial intelligence annals of emergency medicine* (pp. 1192–1199).

Lee, J. D., Li, J. D., Liu, L. C., & Chen, C. M. (2006). A novel driving pattern recognition and status monitoring system. In L.-W. Chang & W.-N. Lie (Eds.), *Advances in image and video technology*, ser. Lecture notes in computer science (Vol. 4319, pp. 504–512). Berlin: Springer.

NHTSA. (2010). http://www.nhtsa.gov

Omidyeganeh, M., Javadtalab, A., & Shirmohammadi, S. (2011). Intelligent driver drowsiness detection through fusion of yawning and eye closure. *IEEE International Conference on Virtual Environments Human-Computer Interfaces and Measurement Systems (VECIMS), 2011*, 1–6.

Sandberg, D., Akerstedt, T., Anund, A., Kecklund, G., & Wahde, M. (2011). Detecting driver sleepiness using optimized nonlinear combinations of sleepiness indicators. *IEEE Transactions on Intelligent Transportation Systems, 12*(1), 97–108.

Saruwatari, K., Sakaue, F., & Sato, J. (2012). Detection of abnormal driving using multiple view geometry in space-time. *IEEE Intelligent Vehicles Symposium (IV), 2012*, 1102–1107.

Zhu, Z., & Ji, Q. (2004). Real time and non-intrusive driver fatigue monitoring. In *Proceedings of 7th international IEEE conference on intelligent transportation systems*, ITSC 2004 (pp. 657–662).

Chapter 2
Road Safety and Intelligent Transport Systems

Abstract This chapter reviews the current state of knowledge regarding safety problems that lead to traffic accidents. It then analyses one potential solution to these safety problems, in the form of ITS, providing a definition and theoretical overview, describing the architecture of ITS and exploring its classification and applications. Finally, the role of Positioning, Navigation and Timing (PNT) is highlighted to show the importance of high-end GNSS technology in transport safety related ITS applications.

2.1 Traffic Accidents and Safety

The increase in traffic in urban areas has been linked to a growing number of accidents and fatalities globally. The relevant statistics and solutions are reviewed below.

2.1.1 Statistics and Analysis of Road Traffic Accidents

Previous studies have shown that human failure or error is the main cause of road traffic accidents, being implicated in about 70 % of accidents. Environmental factors are implicated in about 20 % of accidents, while vehicular factors are identified as probable causes of 10 % (Treat et al. 1980; Aberg and Rimmo 1998).

Direct causes of accidents attributable to human failure include a failure to look properly before manoeuvring, accounting for 40 % of all accidents reported to the police in 2012. Exceeding the speed limit is another frequent human failure implicated in accidents; reported as a factor in 5 % of accidents, although these involved some 14 % of fatalities. Either exceeding the speed limit or travelling too fast for the prevailing conditions was a reported factor in 12 % of all accidents, with these accidents accounting for 24 % of all fatalities. Other human failures such as inattention, inadequate evasive action, and internal distraction also cause accidents.

R. Sun, *An Integrated Solution Based Irregular Driving Detection*,
Springer Theses, DOI 10.1007/978-3-319-44926-5_2

Furthermore, vision (especially poor dynamic visual acuity) and personality (especially poor personal and social adjustment) have been found to be related to the likelihood of an individual being involved in an accident (Treat et al. 1980; Wegman et al. 2001; Kilbey 2013).

In respect to environmental factors, visual obstructions and slippery roads are leading causes of accidents. In many EU Member States, a high proportion of severe accidents involve vehicles leaving the road and collisions with roadside obstacles, especially with trees (Wegman et al. 2001). The main reasons for these types of accidents are bad weather and poor road design.

Vehicular causes of accidents are mainly a result of brake failure, inadequate tread depth, side-by-side brake imbalance, under-inflation and vehicle-related vision obstructions (Treat et al. 1980).

2.1.2 Solutions

Government efforts play a key role in improving road safety. For example, the UK government has set out a strategic framework for road safety with a package of policies to reduce deaths and injuries on the road (Kilbey 2013).

The conventional methods to improve a country's transport system are building new roads or repairing aging infrastructure. Modern transport relies not only on concrete and steel, however, but also on the implementation of technology, specifically a network of sensors and communication devices that collect and disseminate information about the functioning of the transport system. In recent years, many governments have encouraged research into ITS applications in the belief that they can deliver important safety benefits. This research has focused on a wide range of ITS based safety related applications such as real-time traffic alerts, cooperative intersection collision avoidance and on-vehicle systems such as anti-lock braking and lane departure, collision avoidance and crash notification systems. ITS is discussed in detail in Sect. 2.2.

2.2 Intelligent Transport Systems

Intelligent Transport Systems (ITS) are arguably the most efficient technology to solve modern transport problems, enabling more efficient and comfortable transport by providing timely information through communication technologies added to transport infrastructure and vehicles. A key aspect of ITS, therefore, is the use of technology to achieve a tight and active interconnection, both between vehicles themselves, and between vehicles and other aspects of the transport system such as infrastructure, drivers and passengers (Slinn et al. 2005).

Several relevant services, including travel and traffic management, public transport management, electronic payment, commercial vehicle operations, emergency management and advanced vehicle safety systems have been implemented using ITS. An interesting feature of each of these services is the requirement for the relevant ITS architecture to include a positioning or navigation capability. Vehicles equipped with navigation systems can provide accurate position and route planning information for the users, and this information can also be linked to traffic reports. Traffic reporting to the central system and back to other users can be reported in a more timely and accurate manner by ITS. Information about road conditions, such as traffic congestion, detours and accidents, can be passed to the users and other parts of the system in an efficient way in order to influence user decisions on the travel route and thereby to avoid potential traffic problems.

2.2.1 Definition

ITS are road based, vehicle based, vehicle to road based or vehicle to vehicle based technologies supporting the driver and/or the management of traffic in a transport system. The in-vehicle system is an important factor in safety application related ITS applications. There are two major subdivisions on the vehicle side: In-Vehicle Information Systems (IVIS) and Advanced Driver Assistance Systems (ADAS) (Riecken 1997). IVIS and ADAS can each be further subdivided into active and passive safety systems.

Active and passive safety systems can be easily illustrated using the example of the safety belt. The standard safety belt is considered to be a passive system, whereas an intelligent safety system could be a safety belt that automatically tensions when an impact is considered imminent and adapts the tension to the crash severity and the mass of the driver (Ezell 2010).

2.2.2 Architecture

This sub-section describes the ITS architecture. ITS architecture allows ITS enabling technologies to work together as a system, since the effective and secure operation of the entire system, and its interoperability with other systems, is critical.

Generally, the ITS architecture can provide a framework for planning, programming and implementing ITS for the specific country. The architectural framework is comprised of two technical layers, a transport layer and a communication layer, working on an institutional layer (U.S. Department of Transport 2014).

The ITS objectives and requirements, such as institutions, policies, funding mechanisms and processes, are designed at the institutional layer. The transport layer is the key component of ITS architecture, and includes the defined transport solutions for the ITS. These solutions are in the subsystems and interfaces, which contain the underlying functionality and data definitions for each transport service. The communications layer is used to support the transport solutions by providing accurate and timely information exchange between systems. Thus, the ITS architecture is the development of standards of a reference framework for ITS. The following sub-section introduces the applications of ITS.

2.2.3 Classification and Applications

There are different ways of classifying ITS and its two major subdivisions: IVIS and ADAS. Generally, IVIS includes services about route guidance and various vocal/visual information messages, which are provided to the driver; while ADAS includes collision avoidance and vision enhancement, which relates to assistance systems for the primary driving task (Floudas et al. 2004). The ADAS can inform and warn drivers when specific situations occur and provide feedback on their actions.

ITS applications can be applied to different modes of transport on streets and motorways. The users of the system can be private vehicle operators, public transport users and commercial vehicle operators. Advanced Traveller Information Systems (ATIS) and Advanced Transport Management Systems (ATMS) are the main ITS safety related applications.

Advanced Traveller Information Systems (ATIS) provide drivers with real-time travel and traffic information, such as transit routes and schedules, navigation directions and information about delays due to congestion, accidents, weather conditions or road repair work. The most effective traveller information systems are able to inform drivers in real-time of their precise location, inform them of current traffic or road conditions on their own and surrounding roadways, and empower them with optimal route selection and navigation instructions, ideally making this information available on multiple platforms, both in-vehicle and out. The three key facts to the provision of real-time traffic information are collection, processing and dissemination. ATIS also includes in-car navigation systems and telematics-based services, including location-based services, mobile calling, or in-vehicle entertainment options, such as Internet access and music or movie downloads. Other ATIS applications include intelligent parking: in some cities, such as Stockholm and San Francisco, the deployed systems can indicate to drivers the location of available parking spaces in the city, even allowing drivers to reserve parking in advance (Ezell 2010).

Advanced Transport Management Systems (ATMS) include ITS applications that focus on traffic control devices, such as traffic signals, ramp metering, and real-time dynamic message signs on motorways for the driver to receive traffic or motorway status updates. Traffic Operations Centres (TOC) are centralized traffic management centres that collect information from sensors, roadside equipment, vehicle probes, cameras, message signs and other devices to create an integrated view of traffic flow and to detect accidents, dangerous weather events or other roadway hazards. Adaptive traffic signal control refers to dynamically managed, intelligent traffic signal timing, which can be realized by giving traffic signals the capability to detect the presence of waiting vehicles, or giving vehicles the capability to communicate that information to a traffic signal, to improve timing of traffic signals, thereby enhancing traffic flow and reducing congestion. Another ATMS that can provide significant traffic management benefits is ramp metering. Ramp meters are traffic signals on motorway entrance ramps that break up clusters of vehicles entering the motorway, with the function of reducing the disruptions to motorway flow caused by the vehicle clusters, as well as making merging safer (Ezell 2010).

2.3 Role of Positioning, Navigation and Timing

Positioning Navigation and Timing (PNT) is important for ITS. Many advanced aspects of ITS, especially safety related applications, are not possible without position determination capability. This section firstly introduces the requirements of typical safety related ITS applications and then introduces GNSS and other related technologies used for ITS.

2.3.1 Requirements of ITS Applications

The general PNT requirements for ITS applications are specified in terms of the four parameters of accuracy, integrity, continuity and availability as follows (GALA Project 2000; GENESI Project 2007).

Accuracy. Accuracy is used to define the degree of conformance between the estimated position and the true position. Position accuracy performance is expressed as an error in metres with an associated statistical value (e.g. 95 %). The constellation geometrical strength which can be predicted from the satellite ephemeris, the approximate position of the user and the quality of observations, influence the accuracy of the positioning.

Integrity. Integrity is defined as the ability of a system to provide timely warnings to users regarding when the GNSS system should not be used for navigation. There are two types of integrity error. Type one is a false alarm, which rejects a measurement from the true population; Type two is missed detection, which accepts a measurement from the outlier population (GENESI Project 2007). In order to measure the integrity performance, some parameters are used as follows:

Alarm limit. The alarm limit is where the size of the position error exceeds the limit, which the user must be informed of.

Time to Alarm (TTA). TTA is the maximum allowable time between the onset of the error and a warning message reaching the user.

Integrity Risk (per hour). This is the probability that the GNSS terminal provides position information with an error exceeding the alarm limit for a period of longer than the TTA.

Continuity. Continuity is defined as the probability that a system will provide the specified level of navigation accuracy and integrity throughout an operation of a specified period, given that navigation accuracy and integrity are available at the start of the operation.

Availability. Service availability is the percentage of the time that the system is usable, i.e. during which the user terminal provides the required levels of accuracy, integrity and continuity. Unavailability arises from apparently random unpredictable factors.

Based on these parameters, safety related ITS applications require high levels of accuracy and integrity, as discussed below and illustrated in Table 2.1.

Automated Highway: The automated highway application assigns a safe separation distance which should be maintained by subsequent vehicles in the same lane in a highway. This is a very safety critical application which requires high levels of accuracy and integrity. In addition, a large number of users are required for its effective operation and it would therefore have a significant potential market.

Intelligent Speed Assistant: The Intelligent speed assistance application uses Geographic Information Systems (GIS) to determine the speed limit associated with the current section of road. This information may then be relayed to the user in the form of warnings. The liability of such a system must be clear to avoid possible speeding charge counter claims.

Lane Control: Positioning information may be made available to users within a particular lane. Such information could be used in a lane keeping control system to improve safety, especially on motorways, highways and intercity routes (Pohl and Ekmark 2003).

Restraint Deployment: In order to avoid collisions on the highway, high update position and velocity estimations will be used. As a further application, as soon as

Table 2.1 Safety related ITS applications (GENESI Project 2007)

Application name	Accuracy (m)	Integrity			Continuity	Availability (% per 30 days)
		Alarm limit	TTA (s)	Integrity risk		
Automated highway	0.1–1	1 m	1	2×10^{7}	NI	99.8
Intelligent speed assistance	0.01–1	5 m	2	10^{-5}	NI	NI
Lane control/geodetic control	0.01–1	1 m	1	2×10^{-7}	NI	NI
Restraint deployment	1	NI	0.1	NI	NI	99.9

an unavoidable collision is detected, restraints and air bags could be deployed before a crash takes place. This could help to limit injury to the passengers.

Table 2.1 presents the requirements for the safety related ITS applications introduced above, as determined by the GENESI project (GENESI Project 2007). It can be seen that for the safety related applications, which are all based on lane level positioning, sub-metre positioning accuracy is required. In terms of integrity, the requirement varies for different applications. The continuity parameter is not quantified for the safety related applications. Furthermore, availability is not quantified for some applications.

The above requirements and applications are just examples of general information for the positioning system used for the ITS applications. From the table above, accuracy is an important requirement for the PNT system in many ITS applications. Currently, however, there are no specific requirements defined for lane level irregular driving detection systems, despite the fact that this forms an important basis for many safety related ITS applications. Based on the general requirements for the PNT system, the accuracy is developed as the requirement for the precise positioning algorithm of the lane level irregular driving detection system. Integrity is also an important requirement for safety related ITS applications. Currently, however, there are no defined parameters to measure the integrity of the irregular driving detection algorithm. For the safety related ITS applications, it is important to detect the onset of irregular driving promptly and, therefore, the time to first detection has been developed as a parameter for measuring the integrity of irregular driving detection systems. In addition, correct detection rate, which is the percentage of the accurate detection of irregular driving, is also used as a parameter for the integrity, in order to evaluate the performance of irregular driving detection. In addition, availability, which is the percentage of a journey when it is possible to perform a detection, is developed as a requirement for the irregular driving detection. These parameters are quantified in this thesis through simulation and real field data validation.

2.3.2 GNSS for ITS

GNSS positioning is the main high accuracy positioning technique that has been developed in recent years. Currently, there are four main Global Navigation Satellite Positioning Systems (GNSS) either operating or being developed. These are GPS, GLONASS, Galileo and BeiDou. GPS is the most important GNSS system today and has largely the same principle of operation as the other systems. The following discussion is, therefore, mainly focused on GPS.

GPS is the first fully operational GNSS system. The position of the user (and thus the vehicle's position) is calculated from the signals received from several different satellites through GPS receivers in the vehicles' On-Board Units (OBU). The GPS receivers require line of sight to the satellites. Nowadays, GPS is the core technology behind many in-vehicle navigation and route guidance systems. Many countries record miles of automobiles and/or trucks by applying OBU equipped with GPS devices in order to implement user fees based on vehicle miles travelled (Ezell 2010).

Currently, Real Time Kinematic (RTK) and Precise Point Positioning (PPP) are the two advanced techniques that have the potential to achieve sub-metre positioning accuracy for high-end ITS applications. The RTK GPS technique is a relative positioning technique mainly based on carrier phase measurements. Similar to other relative positioning techniques, it employs at least two receivers simultaneously tracking the same satellites, and is suitable for a large number of users/rovers within 15–20 km of a known (reference) point, since the same error sources can be differenced away for the receivers within the specific baseline length. In RTK mode, the base receiver transmits the Radio Technical Communication for Maritime Services (RTCM) format data through a radio link to the rover; the rover receiver can then combine the station's data and its own measurements quickly to calculate the real-time location of the point (Langley 1998). The technique for initial ambiguity resolution used in RTK GPS is the On-The-Fly (OTF) ambiguity resolution technique. This is a fast technique requiring no post-processing and works by instantaneously fixing the integer value of ambiguous parameters by considering a number of possible integer ambiguous values for a number of satellites and searching through them to find the ambiguities which are the most likely to be correct (Hatch1990). The biggest difference between RTK and other relative positioning techniques is that the coordinates of the unknown point can be determined in real-time to centimetre level accuracy provided between the base and rover. The real-time system brings significant advantages, e.g. for establishing positioning in the field and reducing the amount of the data to be archived.

This conventional single baseline approach is largely based on Differential GPS (DGPS), which works on the concept that biases and errors affect both receivers of the same magnitudes within an acceptable baseline length. These errors and biases can therefore, be differenced away by applying appropriate corrections (Alves 2004). Single baseline RTK uses a reference receiver transmitting its raw measurements or observation corrections to a rover receiver via some sort of data communication link (e.g., radio, cellular telephone). The data processing at the rover site includes ambiguity resolution of the differenced carrier phase data and coordinate estimation of the rover position. Provided the maximum distance between the reference and the rover receiver does not exceed 10–20 km, centimetre level accuracy can be obtained. If the distance between the two receivers increases, however, the accuracy will decrease. This limitation is caused by distance-dependent biases such as orbit error and ionospheric error. With a shorter baseline, these errors are differenced away. As the baseline distance increases, however, the error sources for the two receivers become slightly different and, therefore, the observation accuracy decreases (Wanninger 2008).

Network-Based RTK (NRTK) GPS positioning methods overcome the shortcomings of conventional RTK by using a network of reference stations to measure the correlated error over a region and to predict their effects spatially and temporally within the network. Generally, centimetric accuracy can be obtained both in static and kinematic modes. This high level of accuracy is only affected by some level of unavailability caused by GPS signal blockage (Aponte et al. 2008). The main limitation of NRTK is that the initial cost of establishing the positioning infrastructure is expensive, in addition to the maintenance fee for the appropriate density and location of the stations. Moreover, the quality, functionality, integrity and robustness are issues still to be resolved (Rizos et al. 2012).

Precise Point Positioning (PPP) is a recently developed positioning technique, based on the use of correction products for the orbit and clock errors, as well as other error (e.g. ionospheric delay) models to calculate the precise position of the user's receiver antenna with a single GPS receiver, which can be dual or single frequency. Precise satellite orbit and clock information are obtained from the product to correct for the errors in satellite orbits and clocks. Also, sophisticated models for systematic biases are employed. All these methods enable PPP to provide positioning solutions to sub-metre accuracy. Table 2.2 below shows the errors or biases and their mitigation/correction for PPP.

As shown in Table 2.2, most of the errors can be modelled accurately, except for the tropospheric delay and receiver clock errors. Tropospheric delay includes dry

Table 2.2 PPP errors and mitigation

Error sources	Mitigation	Residual range errors after application of mitigation
Satellite orbit errors	Precise orbits and precise orbit models are estimated by IGS products	With IGS final (estimated) products, static residual range error <0.03 m depends on the baseline length
Satellite clock errors	The precise satellite clock is estimated by the IGS products	With IGS final (estimated) products, static residual range error <0.03 m
Satellite antenna phase centre offset and variations	Precise models for satellite antenna phase centre offsets and variations are applied	IGS relative phase centre model, static residual range error <0.01 m
Ionospheric delay	1st order effect mitigated through ionospherically-free observable and higher order effects are modelled	Klobuchar model applied
Tropospheric delay	Requires estimation (absolute effect)	Treated as an unknown parameter in the processing
Receiver clock errors	Requires estimation (absolute effect)	Treated as an unknown parameter in the processing
Receiver antenna phase centre offset and variations	Model can be applied from different manufacturers (absolute effect)	Static residual range error <0.005 m
Station motions	Station motion model can be applied (absolute effect)	Static residual range error <0.01 m

and wet delays, and while the dry delay can be modelled by Hopfield or Sasstamoinen models, the wet delay is still difficult to model. Tropospheric delay and receiver clock errors, therefore, are always treated as unknowns along with the station coordinates and ambiguity parameters. A sequential Least Squares (LSQ) filter is used to process the observation equations and this means the real-time updated observation can be used to improve the positioning solution in a sequential manner. PPP means that users only need one receiver for absolute positioning, which creates greater flexibility and computational efficiency for the user. The limitations of PPP are slow convergence time, and a lack of user equipment to support real-time algorithms and real-time satellite orbit and clock data streams (El-Rabbany 2006; Rizos et al. 2012).

2.4 Summary

In this Chapter, road safety problems and the related solutions have been discussed. ITS, as the most efficient solution to transport safety has been introduced in detail and the positioning for ITS and GNSS for ITS have been discussed. The requirements for safety related applications are illustrated as examples of general information for the positioning system used for the ITS applications. The specific requirements for lane level irregular driving, however, which are the basis for many safety related ITS applications, have hardly been determined by previous research. The accuracy, time to first detection, availability and correct detection rate are requirement parameters used for the system to detect lane level irregular driving. RTK and PPP are two techniques that are able to provide sub-metre positioning accuracy for high-end ITS applications. PPP requires a long convergence time during initialization and is not suitable for highly dynamic applications, while RTK, which exhibits stable positioning accuracy in both static and dynamic modes, is more suitable for lane level irregular driving detection. In the next chapter, the state-of-the-art studies in irregular driving detection will be reviewed.

References

Aberg, L., & Rimmo, P. (1998). Dimensions of aberrant driver behaviour. *Ergonomics, 41*, 39–56.

Alves, P. (2004). Development of two novel carrier phase-based methods for multiple reference station positioning. *Ph.D. Thesis for Department of Geomatics Engineering, the University of Calgary, Calgary.*

Aponte, J., Meng, X., & Burbidge, M. (2007). Performance assessment of a GPS network RTK service. In *2008 Integrating Generations FIG Working Week*, Stockholm.

El-Rabbany, A. (2006). *Introduction to GPS: The Global positioning system (2nd rev. ed.).* Boston: Artech House Publishers. ISBN: 1-59693-016-0.

Ezell, S. (2010). Explaining international IT application leadership: Intelligent transportation systems. In *Information Technology and Innovation Foundation, Washington.*

Floudas, N., Amditis, A., Keinath, A., Bengler, K., & Engeln, A. (2004). Review and taxonomy of IVIS/ADAS applications. *AIDE Deliverable 2.1.2.* http://www.aide-eu.org/pdf/sp2_deliv/aide_d2-1-2_summary.pdf

GALA Project. (2000). Galileo overall architecture definition. Preliminary user requirements. Document ID: GALA-Racal-cs-003, 06/03/2000.

GENESI Project. (2007). GNSS evolution next enhancement of system infrastructures, task 1—Technical note, forecast of mid and long term users' needs. Document ID: ESYS-TN-Task1-001, 08/06/2007.

Hatch, R. (1990). Instantaneous ambiguity resolution. In *KIS-90 Symposium, Banff.*

Kilbey, P. (2013). *Reported Road Casualties in Great Britain: 2012 Annual Report.* Department of Transport.

Langley, R. B. (1998). Communication links for DGPS. *GPS World, 4*(5), 47–51.

Pohl, J., & Ekmark, J. (2003). Development of a haptic intervention system for unintended lane departure. *SAE Technical Paper*, 2003-01-0282, Society of Automotive Engineers.

Riecken, D. (1997). What makes an intelligent user interface intelligent? In *Proceedings of the 2nd international conference on intelligent user interfaces , Orlando, Florida, United States.* New York, NY: ACM Press.

Rizos, C., Janssen, V., Roberts, C., & Grinter, T. (2012). Precise point positioning: Is the era of differential GNSS positioning drawing to an end? In *FIG working week 2012 knowing to manage the territory, protect the environment, evaluate the cultural heritage, Rome, Italy.*

Slinn, M., Matthews, P., & Guest, P. (2005). *Traffic engineering design principles and practice* (2nd ed., pp. 209–217). Burlington, MA: Elsevier Butterworth-Heinemann.

Treat, J. R., Tumbas, N. S., McDonald, S. T., Shinar, D., Hume, R. D., Mayer, R. E., et al. (1980). Tri-level study of the causes of traffic accidents. *Executive Summary*. U.S. Department of Transportation.

U.S. Department of Transport. (2014). http://www.iteris.com/itsarch/

Wanninger, L. (2008). *GPS on the web: Virtual reference stations (VRS).* Retrieved. http://www.wasoft.de/lit/gpswebvrs.html

Wegman, F., Brouwer, M., Cauzard, J. P. & Elvik, R. (2001). *Transport safety performance indicators.* European Transport Safety Council.

Chapter 3
State-of-the-Art in Irregular Driving Detection

Abstract In this chapter, the definition of irregular driving and the importance of its detection are set out, along with related recent research. In particular, for the review of the irregular driving detection research: firstly, the relevant research to date is reviewed; secondly, in-car positioning technology, which is the core technology behind lane level positioning, based on vehicle motion sensors and models and filter algorithms for sensor integration, is presented; finally, driving pattern recognition algorithms are discussed.

In this chapter, the definition of irregular driving and the importance of its detection are set out, along with related recent research. In particular, for the review of the irregular driving detection research: firstly, the relevant research to date is reviewed; secondly, in-car positioning technology, which is the core technology behind lane level positioning, based on vehicle motion sensors and models and filter algorithms for sensor integration, is presented; finally, driving pattern recognition algorithms are discussed.

3.1 Definition of Irregular Driving

Irregular driving behaviour is defined in this thesis as hazardous driving styles, including speeding, weaving, swerving and jerky driving, which can cause traffic accidents when travelling on motorways and urban roads. The function of an irregular driving detection algorithm is to detect different types of irregular driving behaviour at an early stage and to warn the driver, in order to prevent the occurrence of accidents.

R. Sun, *An Integrated Solution Based Irregular Driving Detection*,
Springer Theses, DOI 10.1007/978-3-319-44926-5_3

3.2 Benefits of Irregular Driving Detection

Traffic accidents are a major global problem, both personally for the individuals affected, and financially for the whole of society. The detection of irregular driving, therefore, has the potential to improve transport safety, especially by preventing accidents at an early stage, before lane departure.

Irregular driving behaviour resulting in lane departure where vehicles drift off the road edge or into occupied adjacent lanes is the most common reason for road traffic accidents (Oxley et al. 2004). From the UK Department for Transport database, which contains records of the carriageway on which accidents have occurred, lane departure incidents constitute a high proportion of total accidents on highways (21.1 %) and 'A' class roads (38 %). The side-on impact related to lane departure is one of the main reasons for motorway and 'A' road traffic accidents. Especially, side-on impact accidents at a speed of 60 mph or more, have an high injury rate. Thus, early detection of a vehicle's irregular behaviour within the lane should prevent lane departure and related accidents (DfT 2013).

Table 3.1 and Fig. 3.1 show the reasons for irregular driving. Among all the reasons, the inattention/distraction accounts for 60 % of the accidents, and this category represented the highest percentage of fatalities in the UK in 2012 (DfT 2013).

Inattention/distraction results from fatigued drivers engaged in long drives. Fatigue, even in its earliest stages, can make it difficult for drivers to stay in their lanes. Figure 3.2 shows an example of a sequence of events leading to an accident. The example shows a drowsy driver, weaving and then drifting out of the lane and crashing into vehicles in the other lanes. This type of accident can be avoided or mitigated by deploying countermeasures at different stages. The figure shows the

Table 3.1 Accidents involving people by causal factors in 2012 (DfT 2013)

Causal factor	Persons involved in the accidents in total	Percentage of all persons in the accidents
Inattention/distraction	2956	60.9
Speed/aggressive driving	1470	30.3
Visual impairment/obscuration	308	6.4
Impairment due to alcohol/drugs	251	5.2
Inexperience	220	4.5
Road design/features	190	3.9
Road surface	176	3.6
Other	158	3.3
Vehicle defects	123	2.5

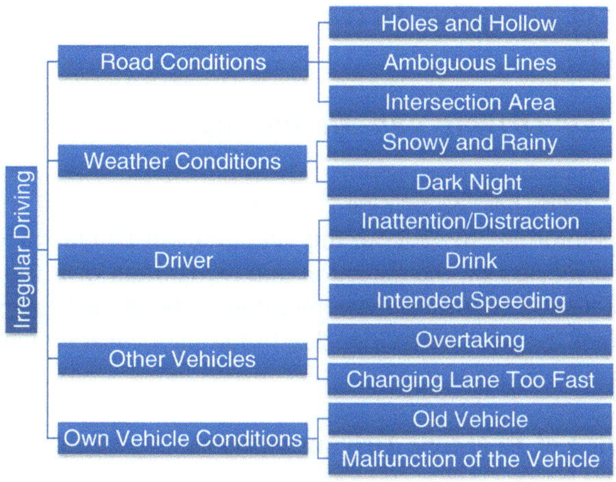

Fig. 3.1 Probable reasons for irregular driving

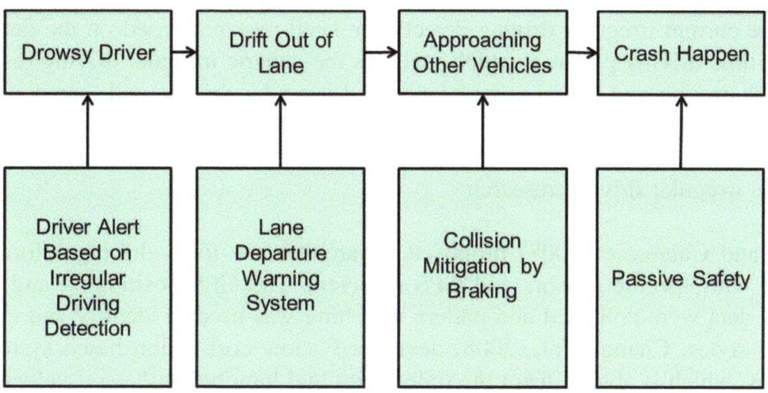

Fig. 3.2 An example of a sequence of events leading to an accident

available techniques for the specific accident sequence. Detection of irregular driving at an early stage recognises the driver's driving pattern and warns when a potential risk exists. This provokes a lane departure warning designed to prevent dangerous lane departure manoeuvers. Finally, collision is mitigated by the automatic application of the brakes just before the accident. It is obvious that the earlier avoidance or mitigation of the accident, the less the potential loss (NHTSA 2010). Thus, detection early stage of irregular driving within the lane is critical to transport safety.

3.3 Review of Irregular Driving Detection Related Research

3.3.1 Irregular Driving and Monitoring

In recent years, there has been a range of research related to irregular driving detection and monitoring. In general, this research can be classified into two categories. The first focuses on the detection and recognition of a vehicle's real-time driving patterns. The second focuses on monitoring the driver's physical behaviour in order to prevent irregular behaviour due to fatigue or inattention.

3.3.1.1 Detection of Real-time Driving Patterns

This category includes the use of sensors for positioning, orientation, velocity and vision to detect vehicle motion data. The data is subsequently analysed to find cues to identify irregular driving and then used for some safety related applications such as collision avoidance or lane departure warnings. The following paragraphs discuss the current irregular driving detection related research based on the detection of real-time driving patterns, which includes the generic irregular driving research, irregular driving and collision avoidance and irregular driving and lane control.

Generic irregular driving research

Lecce and Calabrese (2008) proposed an architecture for a driving information system with specific sensors and GPS receivers. The GPS positioning and acceleration data were collected and pattern matching was used to identify and classify driving styles. Chang et al. (2008) developed a low cost vision based system for vehicles, which is able to learn the trajectories and longitudinal/lateral velocities of the vehicle. Then fuzzy neural networks (FNN) are used according to the image processing information to identify whether the vehicle is about to depart the lane or collide with anything ahead. Krajewski et al. (2009), meanwhile, estimated drivers' fatigue from steering behaviour. The steering data is collected by using orientation sensors and then signal processing is used to capture fatigue impaired patterns. Dai et al. (2010) proposed a system using a mobile phone with an accelerometer and orientation sensor to detect dangerous vehicle manoeuvres typically associated with drunk driving. By computing accelerations based on sensor readings, the user can match the pattern with typical drunk driving patterns extracted from real driving tests to find evidence of drunk driving.

Mohamad et al. (2011) proposed a system for the detection of abnormal driving based on real-time GPS data. The detection algorithm was based on vehicle position and velocity, in particular using maximum speed, maximum acceleration and deceleration as the threshold values. Imkamon et al. (2008) proposed a method for detecting unsafe driving behaviours based on vision and orientation sensors. A fuzzy logic system was used to combine the measured data to classify different levels of hazardous driving. Saruwatari et al. (2012) proposed a method for detecting abnormal driving of vehicles, such as meandering, transverse motion and acceleration or deceleration. In their research, abnormal vehicle motions can be extracted in the sense of group behaviour by using a multi-linear relationship in space-time images. This multi-linear relationship in space-time images could be attained when multiple cameras move with translational motions in different directions at different speeds. Abnormal driving, which is not a translational motion with uniform velocity, can thereby be picked up.

Most of the generic irregular driving research has classified the irregular driving types in a preliminary manner and has not properly quantified the time to first detection, the availability and the correct detection rate of irregular driving. Furthermore, some of the studies used vision sensors where the performance can be affected by the weather conditions. In addition, some studies employing this approach rely significantly on the readings of high grade GPS or other motion sensors, which is not cost effective.

Irregular driving and collision avoidance

Whilst the research reviewed above has focused on the generic detection of irregular driving, the following research works apply the detection of irregular driving to collision-avoidance.

Araki et al. (1996, 1997) developed a collision-avoidance system based on an on-board laser radar and a Charge Coupled Device (CCD) camera, and applied fuzzy logic to evaluate the potential for a collision. The driving status was evaluated by relative distance, relative velocity and the accelerations of both vehicles. Then the risk level of collision was defined based on the comparison with experience data. Risack et al. (2000) developed a lane-keeping assistant video-based system based on vehicle lane position and the time to lane crossing. Their algorithm considered lane geometry and signals from the brakes, steering and indicators. Tsai (2002) also used a laser radar to develop a vehicle safety and warning system and field implementation, while Ueki et al. (2005) developed a vehicular collision-avoidance system by inter-vehicle communication technology. Ferrara and Paderno (2006) presented an automatic pre-crash collision avoidance strategy for cars based on sliding mode control while Tan and Huang (2006) discussed the engineering feasibility of a cooperative collision warning system based on a future trajectory prediction algorithm for vehicles equipped with a differential Global Positioning System (DGPS) unit and other related motion sensors.

Most of the irregular driving and collision avoidance research classifies the risk level in a preliminary way for the collision avoidance application. The research has hardly quantified or even mentioned the performance of different types of irregular driving classification, such as the time to first detection, availability and correct detection rate. Furthermore, most of the research for collision avoidance system uses tightly coupled systems, which compromises compatibility and cost efficiency.

Irregular driving and lane control

This section reviews the research on irregular driving detection related to lane control applications. Kwon et al. (1999) conducted a series of experiments on a vision-based lane departure warning system. The scene in front of the vehicle was captured by a CCD camera. With reference to the steering angle, a perception network approach was applied to build up the decision strategy in order to determine the lane departure of the vehicle. Jung and Kelber (2004, 2005a) developed a Lane Departure Warning System (LDWS) based on a vision system. They designed a linear function to fit the approaching near-vision field, and a quadratic function for the approaching far vision field. They (Jung and Kelber 2005b) subsequently improved their LDWS, by adding the lateral offset, and its rate of change, from video sequences. Lee and Yi (2005) also presented a vision based system for lane-departure detection. Departure ratios are introduced to determine instant lane departure and linear regression to minimize false alarms from noise effects. Yoshida et al. (2006) proposed a lane-departure delay strategy for coordinating responses between the vehicle and the driver. The strategy was tested on a driving simulator.

Most of the irregular driving and lane control research has classified irregular driving types in a preliminary and has not properly quantified the performance of irregular driving. Furthermore, most of the studies used vision sensors, whose performance can be affected by the weather and eliminated lane markings. In addition, some of the research employing this approach relies significantly on high grade sensors and/or other complicated auxiliary systems, which are expensive.

Although the real-time driving pattern detection approach has shown significant potential for the detection of irregular driving, research to date is still at an early stage without credible quantification of performance. Furthermore, a technical barrier that needs to be surmounted is the performance of vision sensor based systems that can be affected by adverse weather conditions. In addition, some research employing this approach relies significantly on high grade GPS and other motion sensors, the cost of which may hinder their widespread practical use. Moreover, most of the above-discussed irregular driving detection systems are still at an early stage of development with neither field tests nor a robust algorithm to distinguish different types of irregular driving styles. Their efficiency and reliability need to be further examined, therefore.

3.3.1.2 Driver Behaviour Monitoring

To date, driver behaviour monitoring has been based either on visual or auxiliary systems to monitor the driver.

Visual observation is an option for detecting driver fatigue. Eriksson and Papanikolopoulos (2001) developed a vision-based approach for diagnosing fatigue by monitoring a driver's eyes and issuing a warning when the system detects irregular eye closure. Zhu and Ji (2004) used two cameras on the dashboard to capture the visual cues of drivers, such as eyelid movement, gaze movement, head movement and facial expression, in order to predict fatigue with a probabilistic model. Lee et al. (2006) used cameras to capture the driver's sight line and driving path and calculated the correlation between them in order to monitor the driving status and patterns. Albu et al. (2008) also used a vision-based system to monitor eye conditions in order to detect fatigue while driving. They claimed that, since sleep onset is the most critical consequence of fatigued driving, it is appropriate to separate the issue of sleep onset from the wider analysis of the physiological state of fatigue simply by judging sleep onset based on the eyes opening and closing. They also installed a force sensor on the accelerator pedal and collected the exerted force to monitor driver fatigue. Omidyeganeh et al. (2011) presented a robust and intelligent scheme for driver drowsiness detection, employing a fusion of eye closure and yawning detection methods. In their approach, the driver's facial appearance is captured via a camera installed in the car and the signs of driver fatigue are studied. Afterwards, the driver state is determined in the fusion phase and a warning message is sent to the driver if drowsiness is detected.

Auxiliary systems coupled with the vehicle to detect interactions between the driver and the vehicle to indicate drunk driving is also an option for monitoring the driver. Heitmann et al. (2001) have proposed various technologies, including a head position sensor, an eye-gaze system, a two pupil-based system and an in-seat vibration system, for alertness monitoring. Based on these technologies, they have used a multi-parametric approach to monitor and prevent driver fatigue. Desai and Haque (2006) assumed that the time derivative of force exerted by the driver at the vehicle-human interface, such as pressure on the accelerator pedal, can be used to decide the level of driver alertness. Sandberg et al. (2011), meanwhile, have presented a method for abnormal driving detection by using physiological signals. For example, signs of sleepiness can be detected by analysing brain activity obtained through electroencephalography. Other auxiliary systems such as electrocardiogram, electromyogram, and skin conductance have also been investigated in the context of detecting driver fatigue and drowsiness (Mohamad et al. 2011).

While monitoring of a driver's physical behaviour has emerged as another potential approach for irregular driving detection, the presence of a vision sensor installed inside the vehicle can lead to a safety hazard due to driver distraction. Moreover, ambiguous road markings and bad weather can seriously affect the performance of the vision sensor. For the auxiliary system, compatibility issues can be raised, since a coupled system is strictly required for system operation. Furthermore, the complexity and high cost of the auxiliary systems can make it difficult to integrate in practice. Most of the research has been based on either simulations or laboratory conditions with very few field trials. Although some research teams have conducted field tests, they generally neither quantify nor address the irregular driving performance. Thus, the driver's physical monitoring based approach to the detection of irregular driving is not yet suited for widespread use.

From the discussion of the two potential approaches for irregular driving detection, it can be concluded that the driver's physical monitoring approach is difficult for real applications, while the driving pattern based irregular driving detection approach is potentially feasible. Currently, there has been hardly any research undertaken to classify the different types of irregular driving methods based on the detection of irregular driving patterns. In addition, most of the current systems are still at an early stage of development and have not quantified the system performance. Having said this, it is acknowledged that these systems could also generally be improved by improved performance of the positioning and orientation detection technology (Krajewski et al. 2009; Dai et al. 2010). In the next section, therefore, in-car positioning technology, which is the underpinning technology for the detection of irregular driving, is reviewed.

3.3.2 In-Car Positioning

One key issue which affects the performance of irregular driving detection systems based on positioning and orientation sensors is the limited access to high accuracy positioning. In this section, the lane level positioning technologies are reviewed, since it is the underpinning technology for irregular driving detection.

Lane level positioning and navigation technology is developed from the current in-car positioning and navigation technology. In order to discuss lane level positioning technology, the main factors for in-car positioning are introduced first. There are basically four different sources of information available for in-car positioning and navigation (Skog and Handel 2009):

1. Various GNSSs and other Radio Frequency (RF)-based navigation systems
2. Motion sensors, radar, visual and laser sensors
3. Road maps
4. Vehicle models and sensor fusion models

The GNSSs receiver and vehicle motion sensors provide observations to estimate the vehicle's state. The vehicle motion models and road maps put constraints on the dynamics of the system and allow past information to be projected forward in time and to be combined with current observational information, whereas the visual, radar and laser sensors can play a major role in the development of ADAS. GNSSs have already been discussed in Chap. 2. Motion sensors and models are introduced in detail in the following subsection.

Table 3.2 Sensors and measurements

Sensor	Measurement
Steering encoder	Front wheel direction
Odometer	Travelled distance
Velocity encoders	Wheel velocities (indirect heading)
Electronic compass	Heading relative to magnetic north
Accelerometer	Acceleration
Gyroscope (rate-gyro)	Angular rotation (velocity)

3.3.2.1 Vehicle Motion Sensors

There are a number of sensors including wheel odometers, magnetometers and accelerometers that can provide information about a vehicle's state that may be used in combination with a GNSS receiver or other absolute positioning system. The most commonly used sensors, together with the information that they provide, are summarized in Table 3.2.

A steering encoder measures the angle of the steering wheel. Hence, it provides a measure of the angle of the front wheels relative to the forward direction of the vehicle platform. Together with information on the wheel speeds of the front wheel pair, the steering angle can be used to calculate the heading rate of the vehicle. Currently, the general angle detection accuracy for the commercial steering encoder varies from ± 3 to ± 20 % (SKF 2015).

An odometer provides information on the travelled curvilinear distance of a vehicle by measuring the number of full and fractional rotations of the vehicle's wheels (Abbott and Powell 1999). This is mainly done by an encoder that outputs an integer number of pulses for each revolution of the wheel. The number of pulses during a time slot is then mapped to an estimate of the travelled distance during the time slot by multiplying by a scale factor, depending on the wheel radius. Non-standard or severely worn tires will therefore result in some errors in the odometer.

A velocity encoder provides a measurement of the vehicle's velocity by observing the rotation rates of the wheels. If separate encoders are used for the left and right wheel of either the rear or front wheel pair, or if separate encoders are used for the wheels on one side of the vehicle, an estimate of the heading change of the vehicle can be found through the difference in wheel speeds. Information on the speed of the different wheels is often available through the sensors used in the Antilock Braking System (ABS) (Carlson et al. 2002, 2004; Hay 2005).

The use of wheel encoders is necessary to form a reliable estimate of the heading changes from the information about the wheel speeds. These notions of how to estimate the travelled distance, velocity and heading of the vehicle are all based on the assumption that wheel revolutions can be translated into linear displacements relative to the ground. In reality, however, there are several sources of inaccuracy or error in the translation of wheel encoder readings to travelled distance, velocity and heading change of the vehicle. These include (Borenstein and Feng 1996):

1. wheel slips
2. uneven road surfaces
3. skidding
4. wheel diameter changes due to variations in temperature, pressure, tread wear and speed
5. wheel diameters unequal between the different wheels
6. efficient wheelbase uncertainties (track width)
7. wheel encoders with limited resolution and sample rate

The first three error sources are non-systematic terrain dependent, which makes it difficult to predict and limit their negative effect on the accuracy of the estimated travelled distance, velocity and heading. The remaining four error sources are systematic, and it is therefore easier to predict their impact on the travelled distance, velocity and heading estimates. In the sensor integration process, the errors due to changes in wheel diameter, unequal wheel diameter, and uncertainties in efficient wheelbase can be reduced by including them as parameters to be estimated. Currently, the claimed accuracy of the velocity encoder is $\pm 10\ \%$ for most manufacturers.

An electronic compass can provide heading measurements relative to the Earth's magnetic north. It is an electronic device, constructed from magnetometers (Abbott and Powell 1999). The declination angle (i.e., the angle between the geographic and magnetic north), which is position dependent, is required to convert the compass heading into an actual north. Thus, knowledge of the compass position is necessary to calculate the heading relative to geographic north.

Generally, the compass is constructed around three magnetoresistive or flux-gate magnetometers, together with pitch and roll sensors (Titterton and Weston 2004). The pitch and roll measurements are needed to determine the attitude between the coordinate system spanned by the magnetic sensors' sensitivity axes and the local horizontal plane, so that the horizontal component of the Earth's magnetic field can be calculated. For a vehicle moving in a planar environment experiencing only small pitch and roll angles, a compass constructed from only two magnetometers with perpendicular sensitivity axes lying approximately in the horizontal plane may be sufficient and cost effective. Power lines and metal structures such as bridges and buildings along the trajectory of the vehicle may cause variations in the local magnetic field, resulting in large and unpredictable errors in the heading estimates of the compass. Magnetic sensors that can be used to detect the vehicle's location with the help of magnets distributed along a highway are also discussed (Yang and Ferral 2003). Currently, the magnetic compass can provide ± 1 degree level of heading accuracy (Honeywell 2015).

An accelerometer provides information about the acceleration of the object to which it is attached. More strictly speaking, an accelerometer produces an output proportional to the specific force exerted on the instrument projected onto the coordinate frame mechanized by the accelerometer (Britting 1971). Information about an object's orientation and rotation may be obtained using a gyroscope (rate-gyro), which measures the angular rotation (velocity) of the object relative to the inertial frame of reference.

Hence, by equipping a vehicle with inertial sensors, i.e., accelerometers and gyroscopes, information about that vehicle's acceleration and rotation is obtained and can be mapped into estimates of the vehicle's attitude, velocity and position. There are many different ways of constructing inertial sensors: a description of common technologies and their typical performance parameters can be found in Titterton and Weston (2004). A description of the trends in inertial sensor technology is offered in Barbour and Schmidt (2001). Historically, due to their high cost, size and power consumption, inertial sensors have mostly been used in high-end navigation systems for missiles, aircraft and marine applications. With the progress in Micro-Electromechanical System (MEMS) sensor technology, however, it has become possible to construct inertial sensors meeting the cost and size demands needed for low-cost commercial electronics, such as vehicle navigation systems.

The errors associated with odometers, velocity encoders and magnetic compasses are partly related to the terrain in which the vehicle is travelling (Titterton and Weston 2004; El-Sheimy and Niu 2007). Unlike these sensors, however, the error sources of inertial sensors are fully self-contained, although these must nonetheless be considered. The most significant inertial sensor errors can be categorized as biases, scale factors, nonlinearities and noise (Titterton and Weston 2004; Grewal et al. 2007).

A bias error is a non-zero output from the sensor for a zero input. Scale factor and nonlinearity errors describe the uncertainty in linear and nonlinear scaling between the input and output, respectively. Random noise exists in the inertial output. Each of these error categories in general includes some or all of the following terms: fixed, turn-on to turn-on variations, random walk, and temperature variations (Grewal et al. 2007).

Calibration of the sensor can estimate and compensate for the fixed terms, such as the temperature varying terms. Turn-on to turn-on terms differ from time to time when the sensor is turned on but stay constant during the operation time, whereas the random walk error slowly varies over time. The turn-on to turn-on and random walk error characteristics of sensors are, therefore, of major concern in the choice of sensors and the information fusion method. In addition to the aforementioned error components, there are also others due to the inevitable imprecision in the mounting of the sensors as well as motion-dependent error components, which may need to be factored into the choice of sensors and information fusion algorithms (Farrell 2007).

To measure the vehicle's dynamics in both long and cross-track directions, a cluster of inertial sensors is needed, which is referred to as an Inertial Sensor Assembly (ISA). Depending on the construction of the navigation system, the ISA may consist solely of accelerometers, but more frequently, a combination of accelerometers and gyroscopes is used (Tan and Park 2005; Chen et al. 1994).

In general, a six-degree-of-freedom ISA (i.e., an Inertial Measurement Unit (IMU) designed for unconstrained navigation in three dimensions) consists of three accelerometers and three gyroscopes, where the sensitivity axes of the accelerometers are mounted to be orthogonal and to span a 3D space, and where gyroscopes measure the rotations around these axes.

Based on the above discussion of the sensors and their measurements, it is clear that choosing the proper sensors for integration with the GPS is critical to the performance of the integrated system. Accelerometers and gyroscopes are two important sensors for the collection of acceleration and yaw rate, respectively. For lane-level irregular driving detection, a vehicle's motion parameters, including high accuracy acceleration, heading angle and yaw rate information are required. Thus, INS is ideal for the integration with GPS, since it contains the accelerometer and gyroscope to measure the acceleration and yaw rate information, which is required for precise positioning estimation.

3.3.2.2 Vehicle Motion Models

From an estimation-theoretical perspective, sensors and vehicle model information play an equivalent role in the estimation of the vehicle state (Julier and Durrant-Whyte 2003). With a perfect vehicle model such that the vehicle state could be perfectly predicted from control inputs, it would be unnecessary to use sensor information. On the contrary, with perfect sensors, the vehicle model would provide no additional information. Neither of these extremes exist in reality, however, and thus it is clear that navigation system performance can be enhanced by utilizing vehicle models. Moreover, the incorporation of a vehicle model in the navigation system may allow the use of less-costly sensors without degradation in navigation performance.

The estimation of a vehicle's dynamic state is one of the most fundamental data fusion tasks for intelligent traffic applications. To achieve this, motion models are applied in order to increase the accuracy and robustness of the estimation. In order to increase the stability and accuracy of the estimation, vehicles are mostly assumed to comply with certain motion models which describe their dynamic behaviour. Another advantage of this approach is the ability to predict the vehicle's position and related dynamic parameters in the future.

The selection of the proper motion model for the process and measurement dynamics of the filtering algorithm is critical. The model should be able to describe the process and measurement dynamics of the system in a framework suitable for the state estimator, as well as provide a proper description of the process and measurement noise statistics. Furthermore, the model must be complete enough to give an adequate description of the system and be sufficiently simple for the filtering algorithm to become computationally feasible.

In the past, numerous motion models (with different degrees of complexity) have been proposed for this task. Some authors have also compared different motion models for certain applications in a rather general way using simulated data (Polychronopoulos 2007; Mattern et al. 2010).

There are many models used to describe the geometric motion of a vehicle. In this thesis, the situation of the vehicle assumption is kinematic, thus only kinematic models are investigated and reviewed. The basic principle of the kinematic motion model is simple, in that it only considers the geometry of the vehicle. The reason for

Fig. 3.3 Graphically shows vehicle geometry

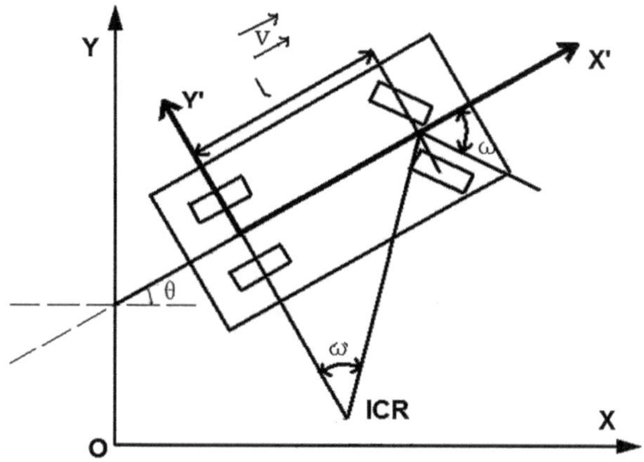

using this assumption is that it is the ideal situation for the kinematic vehicle and it is easy computationally. The assumptions of the design of the basic vehicle model are as follows (Pepy et al. 2006; Ren and Cai 2010; Tham et al. 1998):

1. The vehicle is a rigid solid
2. The axes joining two wheels are parallel to the surface
3. All wheels are perpendicular to the surface without any side deflection
4. The vehicle moves on a planar surface
5. The contact between a wheel and the surface is a point
6. The rotational motion of the wheels is pure roll, without side slip, brake and slide
7. Back wheels can rotate only about a transversal axis of the vehicle
8. The forward wheels turn in a way so as to describe curves centred in their instant rotational centre

Figure 3.3 shows the vehicle geometry. The vehicle has four wheels, the front two are turning wheels and the rear two are driving wheels. XOY is a global coordinate or local coordinate system. X′O′Y′ is the frame of the vehicle body's coordinates. O′ is the middle point of the rear axle of the vehicle, and is the assumed origin of X′O′Y′. l is the distance between the rear and front axles; and ω is the turning angle. Let X_{Local} be the motion posture vector in XOY and X_{Body} be the motion posture vector in X′O′Y′.

$$x_{Local} = [x\, y\, \theta]^T \tag{3.1}$$

Assuming that the vehicle rotates around an Instantaneous Centre of Rotation (ICR) in an ultra-short time, when it changes direction, the instantaneous movement track can be described as a circle, and ρ^r *and* ρ^f are the radii of the ICR circle. The transfer equation from local coordinates to the body frame derivation is shown in (3.2) and (3.3).

$$x_{Body} = Rx_{Local} \tag{3.2}$$

$$R(\theta)^{-1} = \begin{bmatrix} \cos\theta & -\sin\theta & 0 \\ \sin\theta & \cos\theta & 0 \\ 0 & 0 & 1 \end{bmatrix} \tag{3.3}$$

$R(\theta)$ is a transformation matrix which can transform the coordinate of the point in a coordinate frame to a point in another coordinate frame, so that the transfer from body coordinates frame to local coordinates frame can be achieved with the Eq. (3.4).

$$x_{Local} = R(\theta)^{-1} x_{Body} \tag{3.4}$$

The kinematic model based on all these parameters is described in Eq. (3.6) and (3.7). v is defined as the heading velocity of the vehicle. If the rate of the whole vehicle is v_r, the direction along the X′ and v_f is the direction of the yaw angle; the relationship between v_r and v_f is in (3.5).

$$v_f = v_r / \cos\omega \tag{3.5}$$

$$x_{Local} = R(\theta)^{-1} \left[\mathrm{v}\cos\theta \; \mathrm{v}\sin\theta \frac{\mathrm{V_r}}{1} \tan\omega \right]^{\mathrm{T}} \tag{3.6}$$

$$x_{Local} = \left[v\frac{\cos(\theta+\omega)}{\cos\omega} \quad v\frac{\sin(\theta+\omega)}{\cos\omega} \quad \frac{v}{l}\tan\omega \right]^{T} \tag{3.7}$$

There are also some other models derived from the basic model, which can be classified into linear and curvilinear models. Linear motion models assume a Constant Velocity (CV) or a Constant Acceleration (CA) in the time scale of the vehicle motion. Their major advantage is the linearity of the state transition, except for the heading angle included in the transition equation, which allows an optimal propagation of the state probability distribution. Since these models assume straight motions, however, they are not able to take rotations (especially the yaw rate) into account (Bar-Shalom and Li 1995; Richter et al. 2008). Curvilinear models, meanwhile, do take into account the rotations around the z-axis, e.g. the yaw rate. They can be further divided by the state variables, which are assumed to be constant.

The Constant Turn Rate and Velocity (CTRV) model is the simplest model in this model group. CTRV models the movement of the objects tracked by the system at a constant turn rate and constant velocity. It has been used for lane level based vehicle localization which integrates vision and GNSS/INS; commonly in the context of airborne tracking systems. It is also used for vehicle localization using digital maps and coherency images (Mattern et al. 2010).

Assume the state space include these parameters: x, y, which are the local coordinates of the vehicle; θ is the heading angle of the vehicle; v represents the velocity of the vehicle; ω is the yaw rate of the vehicle and T is the time interval between the current and next time epoch. The state space and transition equations for the CTRV model are:

$$x(t) = (x\, y\, \theta\, v\, \omega)^T \tag{3.8}$$

$$x(t+T) = \begin{pmatrix} \frac{v}{\omega}\sin(\omega T+\theta) - \frac{v}{\omega}\sin(\theta) + x(t) \\ -\frac{v}{\omega}\cos(\omega T+\theta) - \frac{v}{\omega}\sin(\theta) + y(t) \\ \omega T + \theta \\ v \\ \omega \end{pmatrix} \tag{3.9}$$

The Constant Turn Rate and Acceleration (CTRA) model assumes that the speed is changing at a constant rate, which is the tangential acceleration, while the object's turn rate also remains constant over time (Schubert et al. 2008).

Compared to the CTRV model, one more parameter, a, which is the acceleration of the vehicle, is added to the CTRA model. The state space and transition equations for the CTRA models are expressed in (3.10) and (3.11):

$$x(t) = (x\, y\, \theta\, v\, a\, \omega)^{\mathrm{T}} \tag{3.10}$$

$$x(t+T) = \begin{pmatrix} \frac{1}{\omega^2}[(v(t)\omega + a\omega T)\sin(\omega \mathrm{T} + \theta(\mathrm{t})) + \mathrm{a}\cos(\theta(\mathrm{t}) + \omega\mathrm{T}) - \mathrm{v}(\mathrm{t})\omega\sin\theta(\mathrm{t}) - \mathrm{a}\cos\theta(\mathrm{t})] + \mathrm{x}(\mathrm{t}) \\ \frac{1}{\omega^2}[(-v(t)\omega - a\omega T)\cos(\omega \mathrm{T} + \theta(\mathrm{t})) + \mathrm{a}\sin(\theta(\mathrm{t}) + \omega\mathrm{T}) + \mathrm{v}(\mathrm{t})\omega\cos\theta(\mathrm{t}) - \mathrm{a}\sin\theta(\mathrm{t})] + \mathrm{x}(\mathrm{t}) \\ \omega T + \theta(t) \\ aT + v(t) \\ a \\ \omega \end{pmatrix} \tag{3.11}$$

More sophisticated models can be created. Some of the models consider the disturbed yaw rate measurements and some can change the yaw angle of the vehicle even if it is not moving. In order to take this problem into account, the correlation between v and ω can be modeled by using the steering angle as a constant variable and deriving the yaw rate from v and ω. The resulting model is called Constant Steering Angle and Velocity (CSAV). A variation on this is the Constant Curvature and Acceleration (CCA) model, in which the velocity is assumed to change linearly (Schubert et al. 2008).

In theory, curvilinear models describe the motion of road vehicles very accurately. Errors may still result from highly dynamic effects, however, such as drifting or skidding (Pepy et al. 2006). Models can be designed to cope with these effects but these will not be considered here for two reasons: firstly, most ITS applications are designed with non-critical dynamics; secondly, the information which is necessary for estimating the additional parameters (e.g. slip from every tire, lateral acceleration) is not observable by sensors.

In general, although more sophisticated models outperform simpler ones, especially in situations where the assumptions of the simple models are not correct, they do not lead to better performance in every case. Furthermore, in most motorway situations, CV, CA, CTRA and CTRV give better results than more sophisticated models such as CCA or CVSA models (Schubert et al. 2008).

3.3.2.3 Sensor Integration

Besides accurate vehicle models, an appropriate sensor integration method is critical. Since the observations and dynamics of the vehicle are always considered as nonlinear processes and measurements, the following discussion on filter algorithms focuses on nonlinear filtering methods, which are used for the sensor integration. This section introduces the two most commonly used filter models for nonlinear estimation: the Extended Kalman Filter (EKF) and the Particle Filter (PF).

The EKF and its variations are developed based on the Kalman Filter (KF). The EKF is similar to the KF but it can be used in nonlinear systems because it linearizes the transformations via the Taylor Expansion. The state transition and observation models of the EKF are not required to be linear functions of the state since it is instead of differentiable functions (Mohinder and Angus 2008).

There are two steps in the EKF process: correction and prediction. In the correction step, the error covariance is computed as it is minimized by the Kalman gain factor. The measurement data is applied to correct state estimation by adding the product of the Kalman gain and the prediction error to the prediction. In the prediction step, the next state is predicted by the current state variables using the system model (Barrios et al. 2006).

The calculation steps for the EKF can be written as follows (Barrios et al. 2006):

1. Calculation of the Kalman Gain

$$K_t = P_t^- H_t^T [H_t P_t^- H_t^T + V_t R_t V_t^T]^{-1} \tag{3.12}$$

2. Correction of the a priori state estimate

$$\hat{x}_t(+) = \hat{x}_t(-) + K_t(z_t - h(\hat{x}_t(-), 0)) \tag{3.13}$$

3. Correction of the a posteriori error covariance matrix estimate

$$P_t(+) = (I - K_t H_t) P_t(-) \tag{3.14}$$

The prediction steps are as follows:

1. Prediction of the state

$$\widehat{x_{t+1}} = f(\hat{x}_t, 0) \tag{3.15}$$

2. Prediction of the error covariance matrix

$$P_{t+1}(-) = A_t P_t(+) A_t^T + W_t Q_t W_t^T \tag{3.16}$$

where

x is the state estimate
z is the measurement data
A is the Jacobian of the system model with respect to the state
W is the Jacobian of the system model with respect to process noise
V is the Jacobian of the measurement model with respect to measurement noise
H is the Jacobian of the measurement model
Q is the process noise covariance
R is the measurement noise covariance
K is the Kalman gain
P is the estimated error covariance
σ is the measurement noise
w processing noise

Theoretically, the main assumption of the EKF is that the system is a linearized dynamic system of KF and a Gaussian distribution is followed by all error terms and measurements. EKF, however, can only be used for systems with a highly nonlinear nature and non-Gaussian noise sources on assumption of Gaussian noise. System performance could be significantly improved by more refined nonlinear filtering approaches, such as Particle Filters (PF), which keep the nonlinear structure of the system (Merwe et al. 2004; Daum 2005).

The PF, also known as a Sequential Monte Carlo (SMC) method, is a sophisticated model estimation technique (Doucet et al. 2001). PF is usually used for estimation of Bayesian models in which the hidden variables are connected in a Markov chain, which is similar to a Hidden Markov Model (HMM) but typically with continuous variables instead of discrete of the latent variables in state space, and are also not restricted to making the exact inference manipulable.

The basic idea of PF is that any Probability Density Functions (PDFs) can be represented as a set of samples (particles), with one set of values for the state variables for each particle. Any arbitrary distribution can be represented by this method, making it optimal for non-Gaussian, multi-modal PDFs. The key advantage of PF is that it uses an approximate representation of any arbitrary PDFs based complex model rather than an exact representation of a simplified Gaussian mode, which is also the key difference between PF and EKF. Besides, with sufficient samples, PF estimates are more approaching the Bayesian optimal estimates. As a result, more accurate estimates can be made with PF than with EKF (Rekleitis 2003). The steps for the processing of particle filters are as follows (Gustafsson et al. 2002).

1. Initialization: generate particles $x_o \sim p_{x_o}, i = 1,\ldots,N$. Each sample of the state vector is referred to as a particle.
2. Measurement update: update the weights by the likelihood: $w_t^i = w_{t-1}^i p(y_t|x_i^t) = w_{t-1}^i p_e(y_t - h(x_i^t)), i = 1,2,\ldots,N$ and normalise to $w_t^i := w_t^i / \sum_i w_t^i$. Then we take $\hat{x} \approx \sum_{i=1}^N w_t^i x_t^i$.
3. Resampling: take N sample with replacement from the set $\{x_i^t\}_{i=1}^N$ where the probability to take sample i is w_t^i. Let $w_t^i = \frac{1}{N}$. This step is also called Sampling Importance Resampling (SIR). Resampling result in the recession of the particles, so the improved method is only resampled when the effective number of samples is less than a threshold N_{th}. So $N_{eff} = \frac{1}{\sum_i (w_t^i)^2} < N_{th};\ 1 \le N_{th} \le N$.
4. Prediction: Take a $f_t^i \sim p_f$, and simulate $x_{t+1}^i = Ax_t^i + B_u u_t + B_f f_t^i$, $i = 1,1,\ldots,N$.
5. Let t: = t + 1 and iterate to item 2.

In summary, the EKF is a sub-optimal approach, since it implements a KF for a system dynamic that results from the linearization of the original non-linear filter dynamics around the previous state estimates. It is dependent on the assumption that the process and sensor noises are Gaussian distributed. The PF, meanwhile, is an optimal approach with the characteristic of being generally applicable and without any requirement for a specific distribution due to the posterior probabilities represented by a set of randomly chosen weighted samples. For the calculation time comparison, EKF usually takes less time to estimate the state, even than PF with the least number of particles. Furthermore, the execution time for PF also increases at a higher rate for a larger numbers of particles (Geng and Wang 2007).

3.3.3 Driving Pattern Recognition

For irregular driving related research, it is critical to define different types of vehicle driving patterns. From the literature, three main approaches are commonly considered for driving pattern recognition. They are neural networks, explicit modelling and pattern matching (Lecce and Calabrese 2008).

A neural network consists of several layers of interconnected nodes (neurons). Through a learning process where layers and interconnections are rearranged, an adequate system model is then developed based on the trained neural network. Designing a specific system model is not required (Zhao 1997). Neural networks and explicit modelling represent two supervised ways to teach a system the way to follow a certain objective.

An appropriate set of training samples are required by the neural network and the explicit modelling requires many parameters to be tuned in order to get valid results. The collection of large samples is required by both neural networks and explicit modelling, otherwise the results' error will be affected by the samples.

Pattern matching is based on a pre-built knowledge of some characteristics of the combination of different parameter ranges for different types of driving patterns. Detected parameters are then matched in a real-time system state to check if they correspond to one of the predefined ranges in order to judge the driving style. The Fuzzy Inference System (FIS) is a widely used pattern matching method for detecting driver behaviour. The FIS is an inference system that maps input to output using fuzzy logic through combination rules. Compared to traditional logic theory, where binary sets have two values: true or false, fuzzy logic variables define a truth value that ranges in degree between 0 and 1. Fuzzy logic therefore extends traditional logic theory to deal with the concept of partial truth, where the truth value is in a range between completely true to completely false. Fuzzy logic inference is a simple approach to solving problems instead of attempting to model it mathematically, which results in the FIS depending on human experience more than the technical understanding of the problem (Lecce and Calabrese 2008).

FIS have three stages (Aljaafreh 2012):

1. Fuzzification: mapping of any input to a degree of membership in one or more membership functions; the linguistic condition will be used to evaluate the input variable.
2. Fuzzy inference: calculate fuzzy output.
3. Defuzzification: conversion of the fuzzy output to a crisp output.

In summary, neural networks and explicit modelling, which have the disadvantage of requiring large samples for the operation of the algorithm, are not suitable for irregular driving detection. FIS based pattern matching, however, which does not require large samples for algorithm operation, is suitable for driving pattern recognition.

3.4 Summary

The definition and benefits of irregular driving detection have been presented in this chapter, followed by an overview of research related to the monitoring and detection of irregular driving. Also, in-car positioning technology, which is the core technology behind the detection of irregular driving, has been addressed, particularly with respect to vehicle motion sensors, vehicle motion models and sensor integration. Finally, driving pattern recognition methods have been reviewed and the FIS, with the merit of not requiring large samples for the operation of the algorithm is shown to be more suitable for driving pattern recognition than neural networks and explicit modelling. Thus, real-time vehicle driving pattern recognition based on integrated high accuracy positioning with FIS algorithms reviewed in this chapter are used to develop the system for lane level irregular driving detection in Chap. 4.

References

Abbott, E., & Powell, D. (1999). Land-vehicle navigation using GPS. *Proceedings of the IEEE, 87* (1), 145–162.

Albu, A. B., Widsten, B., Wang, T., Lan, J., & Mah, I. (2008). A computer vision-based system for real-time detection of sleep onset in fatigued drivers. *IEEE Intelligent Vehicles Symposium, 2008*, 25–30.

Aljaafreh, A. (2012). Web driving performance monitoring system. *World Academy of Science, Engineering and Technology*, 70.

Araki, H., Yamada, K., Hiroshima, Y., & Ito, T. (1996). Development of rear-end collision avoidance system. In *Proceedings of IEEE intelligent vehicle symposium* (pp. 224–229).

Araki, H., Yamada, K., Hiroshima, Y., & Ito, T. (1997). Development of rear-end collision avoidance system. *JSAE Review, 18*(3), 314–316.

Barbour, N., & Schmidt, G. (2001). Inertial sensor technology trends. *IEEE Sensors Journal, 1*(4), 332–339.

Barrios, C., Himberg, H., Motai, Y., & Sadek, A. (2006). Multiple model framework of adaptive extended kalman filtering for predicting vehicle location. *Proceedings of IEEE ITSC, 2006*, 1053–1059.

Bar-Shalom, Y., & Li, X. R. (1995). *Multitarget-multisensor tracking: Principles and techniques*. Storrs, CT: YBS Publishing.

Borenstein, J., & Feng, L. (1996). Measurement and correction of systematic odometry errors in mobile robots. *IEEE Transactions on Robotics and Automation, 12*(6), 869–880.

Britting, K. (1971). *Inertial navigation systems analysis*. New York: Wiley.

Carlson, R., Gerdes, J., & Powell, J. (2002). Practical position and yaw rate estimation with GPS and differential wheelspeeds. In *Proceedings of 6th international symposium on advanced vehicle control (AVEC), Hiroshima, Japan.*

Carlson, R., Gerdes, J., & Powell, J. (2004). Error sources when land vehicle dead reckoning with differential wheelspeeds. *Navigation, 51*(1), 13.

Chang, T. H., Hsu, C. S., Wang, C., & Yang, L. K. (2008). Onboard measurement and warning module for irregular vehicle behaviour. *IEEE Transactions on Intelligent Transportation Systems, 9*(3), 501–513.

Chen, J., Lee, S., & DeBra, D. (1994). Gyroscope free strapdown inertial measurement unit by six linear accelerometers. *Journal of Guidance, Control and Dynamics, 17*(2), 286–290.

Dai, J., Teng, J., Bai, X., Shen, Z., & Xuan, D. (2010). Mobile phone based drunk driving detection. In *2010 4th international conference on pervasive computing technologies for healthcare (pervasive health)* (p. 1).

Daum, F. (2005). Nonlinear filters: Beyond the Kalman filter. *IEEE Aerospace and Electronic Systems Magazine, 20*(8), 57–69.

Desai, A. V., & Haque, M. A. (2006). Vigilance monitoring for operator safety: A simulation study on highway driving. *Journal of Safety Research, 37*(2), 139–147.

DfT. (2013). Reported road casualties in Great Britain: Main results 2012.

Doucet, A., De Freitas, N., & Gordon, N. J. (2001). *Sequential Monte Carlo methods in practice*. New York: Springer.

El-Sheimy, N., & Niu, X. (2007). The promise of MEMS to the navigation community. *Inside GNSS, 2*(2), 46.

Eriksson, M., & Papanikolopoulos, N. P. (2001). Driver fatigue: A vision-based approach to automatic diagnosis. *Transportation Research Part C: Emerging Technologies, 9*(6), 399–413.

Farrell, J. (2007). Inertial instrument error characterization. *Navigation, 53*(3), 169.

Ferrara, A., & Paderno, J. (2006). Application of switching control for automatic pre-crash collision avoidance in cars. *Nonlinear Dynamics, 46*(3), 307.

Geng, Y., & Wang, J. (2007). Adaptive estimation of multiple fading factors in Kalman filter for navigation applications. *GPS Solutions, 12*(4), 237–279.

Grewal, M., Weill, L., & Andrews, A. (2007). *Global positioning systems, inertial navigation and integration* (2nd ed.). Hoboken, NJ: Wiley.

Gustafsson, F., Gunnarsson, F., Bergman, N., Forssell, U., Jansson, J., Karlsson, R., et al. (2002). Particle filters for positioning, navigation and tracking. *IEEE Transactions on Signal Processing, 50*, 425–435.

Hay, C. (2005). Turn, turn, turn. *GPS World, 16*(10), 37.

Heitmann, A., Cuttkuhn, R., Aguirre, A., Trutschel, U., & Moore-Ede, M. (2001). Technologies for the monitoring and prevention of driver fatigue. In *Proceedings of the fifth international driving symposium on human factors in driver assessment, training and vehicle design* (pp. 81–86).

Honeywell. (2015). http://honeywell.com

Imkamon, T., Saensom, P., Tangamchit, P., & Pongpaibool, P. (2008). Detection of hazardous driving behavior using fuzzy logic. In *5th international conference on electrical engineering/electronics, computer, telecommunications and information technology, ECTI-CON 2008* (Vol. 2, p. 657).

Julier, S., & Durrant-Whyte, H. (2003). On the role of process models in autonomous land vehicle navigation systems. *IEEE Transactions on Robotics and Automation, 19*(1), 1–14.

Jung, C. R., & Kelber, C. R. (2004). A lane departure warning system based on a linear-parabolic lane model. In *Proceedings of IEEE intelligent vehicles symposium* (pp. 891–895).

Jung, C. R., & Kelber, C. R. (2005a). Lane following and lane departure using a linear parabolic model. *Image and Vision Computing, 23*(13), 1192–1202.

Jung, C. R., & Kelber, C. R. (2005b). A lane departure warning system using lateral offset with uncalibrated camera. In *Proceedings of 8th international IEEE conference on intelligent transportation systems* (pp. 102–107).

Krajewski, J., Sommer, D., Trutschel, U., Edwards, D., & Golz, M. (2009). Steering wheel behavior based estimation of fatigue. In *Proceedings of the fifth international driving symposium on human factors in driver assessment, training and vehicle design* (pp. 118–124).

Kwon, W., Lee, J. W., Shin, D. K., Roh, D., Kim, Y., & Lee, S. (1999). Experiments on decision making strategies for a lane departure warning system. In *Proceedings of IEEE international conference on robotics and automation* (pp. 2596–2601).

Lecce, V. D., & Calabrese, M. (2008). Experimental system to support real time driving pattern recognition. In *Advanced intelligent computing theories and applications with aspects of artificial intelligence annals of emergency medicine* (pp. 1192–1199).

Lee, J. D., Li, J. D., Liu, L. C., & Chen, C. M. (2006). A novel driving pattern recognition and status monitoring system. Lecture notes in computer science. In L.-W. Chang & W.-N. Lie (Eds.), *Advances in image and video technology* (Vol. 4319, pp. 504–512). Berlin: Springer.

Lee, J. W., & Yi, U. K. (2005). A lane-Departure Identification Based on LBPE, Hough Transform, and Linear Regression. *Computer Vision and Image Understanding, 99*(3), 359–383.

Mattern, N., Schubert, R., & Wanielik, G. (2010). High-accurate vehicle localization using digital maps and coherency images. In *2010 IEEE intelligent vehicles symposium*. San Diego, CA: University of California.

Merwe, R., Wan, E., & Julier, S. (2004). Sigma-point Kalman filters for nonlinear estimation and sensor-fusion: Applications to integrated navigation. Presented at the AIAA guidance, navigation, and control conference and exhibit, Providence, RI.

Mohamad, I., Ali, M., & Ismail, M. (2011). Abnormal driving detection using real time global positioning system data. In *2011 IEEE international conference on space science and communication (iconspace)* (p. 1).

Mohinder, S. G., & Angus, P. A. (2008). *Kalman filtering: Theory and practice using MATLAB*. New York: Wiley-IEEE Press. ISBN 978-0-470-17366-4.

NHTSA. (2010). http://www.nhtsa.gov

Omidyeganeh, M., Javadtalab, A., & Shirmohammadi, S. (2011). Intelligent driver drowsiness detection through fusion of yawning and eye closure. *IEEE International Conference on*

Virtual Environments Human-Computer Interfaces and Measurement Systems (VECIMS), 2011, 1–6.

Oxley, J., Corben, B., Koppel, S., Fildes, B., Jacques, N., Symmons, M., et al. (2004). *Cost-effective infrastructure measures on rural roads. Report no. 217*. UK Department for Transport.

Pepy, R., Lambert, A., & Mounier, H. (2006). Reducing navigation errors by planning with realistic vehicle model. *IEEE Intelligent Vehicles Symposium, 2006*, 300–307.

Polychronopoulos, A. (2007). Sensor fusion for predicting vehicles' path for collision avoidance systems. *IEEE Transactions on Intelligent Transportation Systems, 8*(3), 72.

Rekleitis, I. (2003). A particle filter tutorial for mobile robot localization. In *International conference on robotics and automation (ICRA), 2003*.

Ren, X., & Cai, Z. (2010). Kinematics model of unmanned driving vehicle. In *Proceedings of the 8th world congress on intelligent control and automation, Jinan, China*.

Richter, E., Schubert, R., & Wanielik, G. (2008). Advanced filtering techniques for multisensor vehicle tracking. In *Proceedings of IEEE intelligent vehicle symposium*.

Risack, R., Mohler, N., & Enkelmann, W. (2000). A video-based lane keeping assistant. In *Proceedings of IEEE intelligent vehicles symposium* (pp. 356–361).

Sandberg, D., Akerstedt, T., Anund, A., Kecklund, G., & Wahde, M. (2011). Detecting driver sleepiness using optimized nonlinear combinations of sleepiness indicators. *IEEE Transactions on Intelligent Transportation Systems, 12*(1), 97–108.

Saruwatari, K., Sakaue, F., & Sato, J. (2012). Detection of abnormal driving using multiple view geometry in space-time. *IEEE Intelligent Vehicles Symposium (IV), 2012*, 1102–1107.

Schubert, R., Richter, E., & Wanielik, G. (2008). Comparison and evaluation of advanced motion models for vehicle tracking. In *2008 11th international conference on information fusion* (pp. 1–6).

SKF Corporation. (2015). http://www.skf.com

Skog, I., & Handel, P. (2009). In-car positioning and navigation—A survey. IEEE Trans. Intell. Transp. Syst, 10, 4–21, doi: 10.1109/TITS.2008.2011712.

Tan, H. S., & Huang, J. H. (2006). DGPS-based vehicle-to-vehicle cooperative collision warning: Engineering feasibility viewpoints. *IEEE Transactions on Intelligent Transportation Systems, 7*(4), 415.

Tan, C., & Park, S. (2005). Design of accelerometer-based inertial navigation systems. *IEEE Transactions on Instrumentation and Measurement, 54*(6), 2520–2530.

Tham, Y. K., Wang, H., & Teoh, E. K. (1998). Adaptive state estimation for 4-wheel steerable industrial vehicles. In *Proceedings of 37th IEEE conference on decision and control, Tampa, FL*.

Titterton, D., & Weston, J. (2004). *Strapdown inertial navigation technology* (2nd ed.). Washington, DC: AIAA.

Tsai, D. H. (2002). Vehicle safety and collision warning system. In *Proceedings of 5th Asia-Pacific ITS forum, Seoul, Korea*.

Ueki, J., Mori, J., Nakamura, Y., Horii, Y., & Okada, H. (2005). Vehicular collision avoidance support system (VCASS) by inter-vehicle communications for advanced ITS. *IEICE Transactions on Fundamentals of Electronics, Communications and Computer Science, 88* (7), 1816–1823.

Yang, Y., & Farrell, J. (2003). Magnetometer and differential carrier phase GPS-aided INS for advanced vehicle control. *IEEE Transactions on Robotics and Automation, 19*(2), 269.

Yoshida, H., Mouri, H., Sato, S., & Nagai, M. (2006). A study on vehicle lane departure delay system. *Review of Automotive Engineering, 27*(2), 273.

Zhao, Y. (1997). *Vehicle location and navigation systems*. Norwood, MA: Artech House.

Zhu, Z., & Ji, Q. (2004). Real time and non-intrusive driver fatigue monitoring. In *Proceedings of 7th international IEEE conference on intelligent transportation systems*, ITSC 2004 (pp. 657–662).

Chapter 4
A New System for Lane Level Irregular Driving Detection

Abstract In this chapter, a new system for the detection of lane level irregular driving is presented. In Sect. 4.1, an overview of the system is presented. Sections 4.2 and 4.3 present in detail the design of a lane level positioning system with a precise vehicle motion model and PF/EKF filters based, on both straight and curved roads. Section 4.4 details the fuzzy inference system design for different types of irregular driving detection.

In the previous chapters, the state-of-the-art in the monitoring and identification of irregular driving has been discussed. In this chapter, a new system for the detection of lane level irregular driving is presented. In Sect. 4.1, an overview of the system is presented. Sections 4.2 and 4.3 present in detail the design of a lane level positioning system with a precise vehicle motion model and PF/EKF filters based, on both straight and curved roads. Section 4.4 details the fuzzy inference system (FIS) design for different types of irregular driving detection.

4.1 System Overview

The system is designed to link high precision positioning estimation with the different types of irregular driving detection. The estimation of the precise vehicle positioning and dynamic parameters is achieved by integration of precise vehicle models with GPS/INS based positioning measurement, subsequently used with an irregular driving detection algorithm to identify different types of irregular driving patterns.

The requirements for the system can be divided into two parts as discussed in Chap. 2. Accuracy is the parameter used to evaluate the designed lane level positioning algorithm. Time to first detection, availability and correct detection rate are three parameters used to evaluate the driving classification algorithm. The time to first detection is the first time a specific type of irregular driving is detected.

R. Sun, *An Integrated Solution Based Irregular Driving Detection*,
Springer Theses, DOI 10.1007/978-3-319-44926-5_4

The system is designed based on specific irregular scenarios. The most common irregular driving styles for motorways are weaving, swerving and jerky driving on the straight lane. On the curved lane, weaving and jerky driving can also occur, but swerving takes the form of over-turning or under-turning (NHTSA 2010). Figure 4.1 presents the different types of driving styles.

Based on the defined scenarios, *omega* and *d*, which are the vehicle's yaw rate and lateral displacement respectively, represent the vehicle's manoeuvres. In order to smooth the noises of the filter estimated values and extract the trend of their changes, the Moving Average Deviation (MAD) of *omega* and *d*, noted as O-indicator and D-indicator respectively, are developed to represent the driving features of the vehicle. The system to detect irregular driving has to recognize a

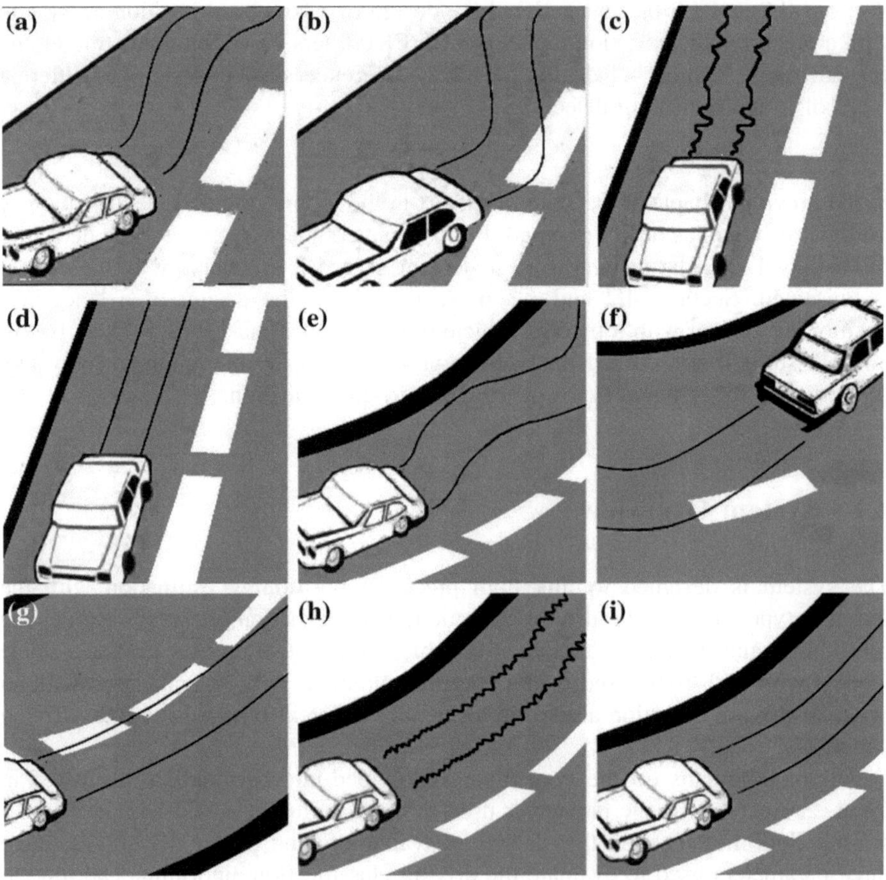

Fig. 4.1 Driving styles **a** scenario 1 weaving on straight, **b** scenario 2 swerving on straight, **c** scenario 3 jerky driving on straight, **d** scenario 4 normal driving on straight, **e** scenario 5 weaving on curve, **f** Scenario 6a swerving (over-turning) on curve, **g** scenario 6b swerving (under-turning) on curve, **h** scenario 7 jerky driving on curve and **i** scenario 8 normal driving on curve

driving event that characterizes the different driving styles based on the irregular driving detection algorithms applied on the O-indicator and D-indicator derived from the filter estimated *omega* and *d* at every time epoch.

The Framework of the system in Fig. 4.2 shows the designed lane level irregular driving detection system, which contains two main parts. The first part is a lane level precise positioning and parameter estimation algorithm. In order to collect the vehicle's driving data, one INS with one gyro and one accelerometer mounted along the vehicle body axis is used to output the yaw rate, acceleration and heading angle for the vehicle heading direction. One GPS receiver is used to determine the vehicle's local coordinates and heading velocity. The initial position is used to feed the PF/EKF models with the precise vehicle motion models to provide the estimated positioning and attitude parameters for the next time epoch for iteration. The second

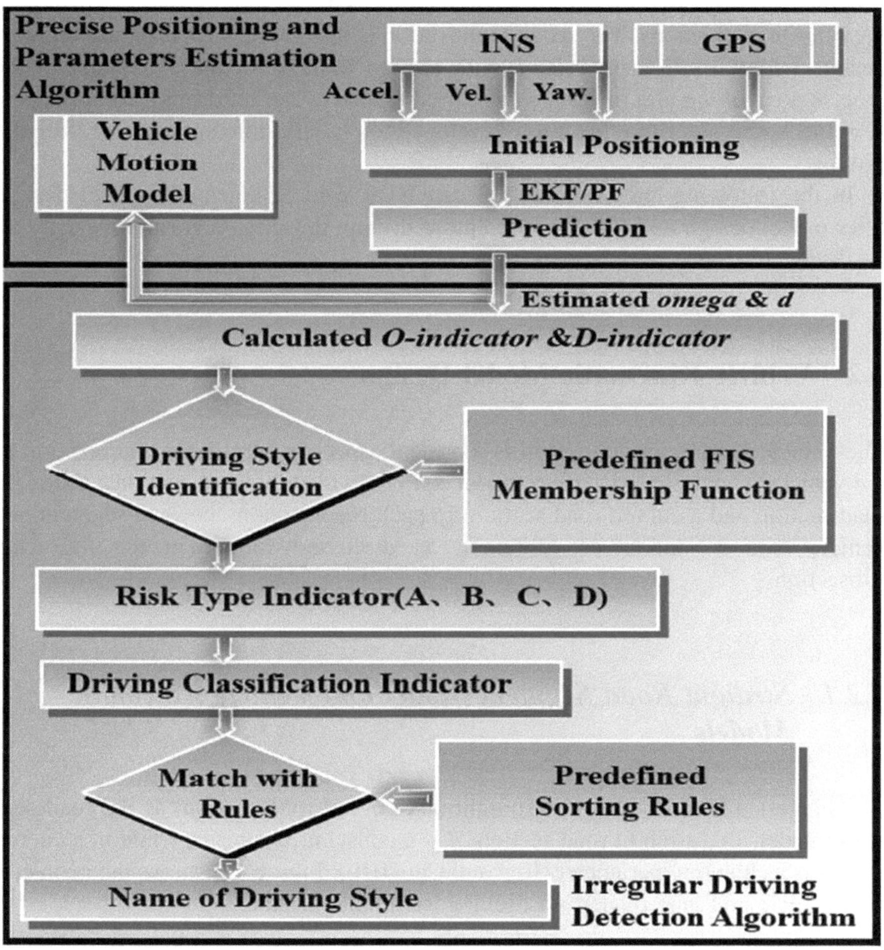

Fig. 4.2 Framework of a new system for the identification of lane level irregular driving

part is the irregular driving detection algorithm. This starts with the calculation of the O-indicator and D-indicator based on the estimated *omega* and *d* from the first part. These two indicators are then used as the input of the FIS for the driving pattern identification, allowing the FIS to output the risk type indicator. In order to amplify the features of each driving style, the driving classification indicators are then developed based on the risk type indicators. Finally, by comparing the sorting of the calculated driving classification indicators with predefined sorting rules extracted from the reference data, the system will output the identified driving style type, including weaving, swerving, jerky driving and normal driving on straight and curved lanes.

The innovations in the designed system are illustrated as follows. The first innovation is that the proposed method takes advantage of merging high accuracy vehicle positioning and dynamic parameter estimations with different types of irregular driving identification technology, which has not been considered by previous investigations. The second innovation is that the new EKF/PF algorithm is designed for GPS/INS integration with motion models for the estimation of the precise positioning and the *omega* and *d* parameters. The third innovation is that a new FIS based algorithm has been developed to classify different types of irregular driving.

In the following sub-sections, the details of vehicle kinematic model design, filter model design and FIS based irregular driving detection algorithm design will be discussed.

4.2 Vehicle Kinematic Model Design

The vehicle motion model design is a critical aspect in identifying the position of the vehicle. The designed system model can be divided into two sections: a straight road section and a curved road section. In each road section, the road segment and vehicle motion models are different, as discussed further in the following subsection.

4.2.1 Straight Road Segment Model and Vehicle Kinematic Models

The biggest difference between straight and curved road sections is the road segment model. In a straight road section, β is a constant parameter, while in a curved road it is a changing parameter (Ren and Cai 2010). Figure 4.3 shows the geometry of a vehicle and lane related positioning on a straight lane.

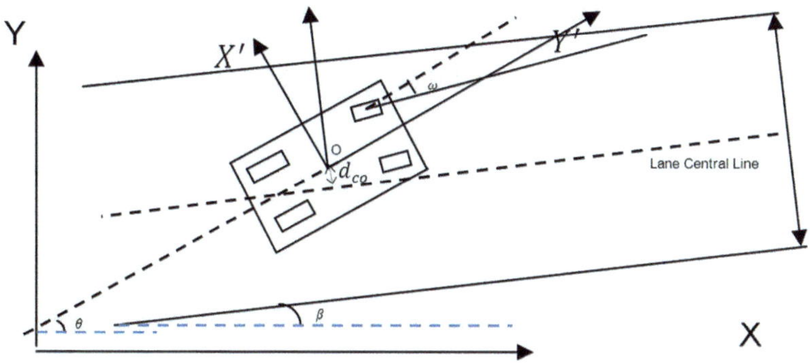

Fig. 4.3 Geometric relationship between vehicle motion model and lane on a straight lane. Where X, Y are the local UK Ordnance Survey National Grid coordinates; X', Y' are the vehicle body frame coordinates; o is the centre of the vehicle; θ is the angle between the vehicle body frame and the local UK National Grid coordinate frame; ω is the yaw rate; d_{co} is the distance between point o (body centre) and the lane central line; β is the angle between the lane segment and the local UK National Grid coordinates

In order to connect the vehicle velocity and positioning with the lane geometry, a state vector of the vehicle motion model is defined in (4.1) (Toledo-Moreo et al. 2009).

$$x(t) = (x\, y\, v\, \theta\, \omega\, a\, \beta\, d_{co})^T \tag{4.1}$$

x, is the X-axis coordinate (in metres) of point o in the local UK National Grid coordinate system; y, is the Y-axis coordinate (in metres) of point o in the local UK National Grid coordinate system; v, is the velocity at the heading direction (m/s); θ, is the heading angle of the vehicle (rad); ω, is the vehicle yaw rate (rad/s); a, is heading acceleration of the vehicle ($\mathrm{m/s^2}$); β, is the angle between the lane and the local UK National Grid coordinates (rad); d_{co}, is the distance between point o (body centre) and the lane central line, which is the vehicle lateral displacement (m).

Different vehicle motion models can create different vehicle state transition models. In the straight road section, on the motorway and in a very short time period (i.e., 0.1 s), the vehicle motion can be assumed to be of constant velocity (CV) or constant acceleration (CA).

The assumption of CV and CA models imply linear motion models. As discussed in Chap. 3, the linearity of the state transition equation is the major advantage of the models but the disadvantage is that the motions have to be assumed as straight and, therefore, rotations, especially the yaw rate are not taken into account (Schubert et al. 2008).

The vehicle state transition model can be defined based on the assumptions of CV and that, in the initial position, the vehicle central line corresponds with the lane central line. The transition equations for the CV model can be expressed as follows.

$$x_{t+1} = x_t + \cos(\theta_t) v_t T \tag{4.2}$$

$$y_{t+1} = y_t + \sin(\theta_t) v_t T \tag{4.3}$$

$$v_{t+1} = v_t \tag{4.4}$$

$$\theta_{t+1} = \theta_t \tag{4.5}$$

$$a = 0 \tag{4.6}$$

$$\beta = constant \tag{4.7}$$

The vehicle state transition model can also be defined based on the assumption of CA, in which case the kinematic model can be expressed as:

$$x_{t+1} = x_t + \cos(\theta_t) v_t T + \frac{1}{2} a_t T^2 \tag{4.8}$$

$$y_{t+1} = y_t + \sin(\theta_t) v_t T + \frac{1}{2} a_t T^2 \tag{4.9}$$

$$v_{t+1} = v_t + a_t T \tag{4.10}$$

$$\theta_{t+1} = \theta_t \tag{4.11}$$

$$a_{t+1} = a_t \tag{4.12}$$

$$\beta = constant \tag{4.13}$$

After developing the prediction parameters in either model, d_{co} is then calculated by the point to the point distance formula. (x_c, y_c) are the points on the central line. d_{co} should comply with the shortest distance between (x_c, y_c) and (x_t, y_t).

$$d_{co} = \min \sqrt{(x_t - x_c)^\wedge 2 + (y_t - y_c)^\wedge 2} \tag{4.14}$$

x_t, is the X-axis coordinate (in metres) of point o in the local UK National Grid coordinate system in time epoch t; y_t, is the Y-axis coordinate (in metres) of point o in the local UK National Grid coordinate system in time epoch t; x_c, is the X-axis coordinate (in metres) of the point on the lane central line in the local UK National Grid coordinate system in time epoch t; y_c, is the Y-axis coordinate (in metres) of the point on the lane central line in the local UK National Grid coordinate system in time epoch t.

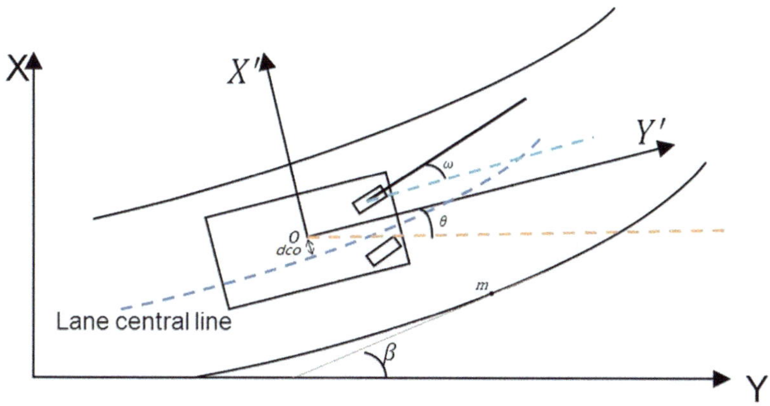

Fig. 4.4 Geometric relationship between vehicle motion model and lane on a curved lane. Where X, Y are the local UK National Grid coordinates; X', Y' are the vehicle body frame coordinates; o is the centre of the vehicle; θ is the angle between the vehicle body frame and the local UK National Grid coordinate frame; ω is the yaw rate; d_{co}, is the distance between point o (body centre) and the lane central line; β is the angle between the lane segment and the local UK National Grid coordinates

4.2.2 Curved Road Segment Model and Vehicle Kinematic Models

On a curved road, the vehicle and lane related positioning can be defined as in Fig. 4.4.

A curved road can be divided into small segments approximating straight sections. On a curved road, for any point m on the curve, β is changing all the time. To calculate d_{co} the algorithm can search the nearest two points on the lane central line and define the new straight line of the two points then calculate the distance of o to the defined straight line. The calculation distance from the central point o to the straight line is the same with the straight road segment model, see Eq. (4.14).

In the case of a curved road, the vehicle motion can be assumed as CTRV or CTRA. Both kinematic models have different transition models, illustrated as follows.

The state vector of the vehicle motion model can be defined as:

$$x(t) = (x\; y\; v\; \theta\; \omega\; a\; \beta\; d_{co})^T \tag{4.15}$$

x, is the X-axis coordinate (in metres) of point o in the local UK National Grid coordinate system; y, is the Y-axis coordinate (in metres) of point o in the local UK National Grid coordinate system; v, is the heading velocity of the vehicle (m/s); θ, is the heading angle of the vehicle (rad); ω, is the vehicle yaw rate (rad/s); a, is the acceleration of the vehicle (a = 0 in CTRV model) $(\mathrm{m/s^2})$; β, is the angle between the lane and the local UK National Grid coordinate (rad); d_{co}, is the shortest distance between point O to the corresponding tangent of the central line of the lane (m).

In the CTRV model, the vehicle state transition can be expressed as,

$$\begin{pmatrix} x_{t+1} \\ y_{t+1} \\ v_{t+1} \\ \theta_{t+1} \\ \omega_{t+1} \end{pmatrix} = \begin{pmatrix} x_t \\ y_t \\ v_t \\ \theta_t \\ \omega_t \end{pmatrix} + \begin{pmatrix} \frac{v_t}{\omega_t}(\sin(\omega_t T + \theta_t) - \sin\theta_t) \\ \frac{v_t}{\omega_t}(-\cos(\omega_t T + \theta_t) + \cos\theta_t) \\ v_t \sin(\theta_t + \omega T - \beta_t) \\ \omega_t T \\ 0 \end{pmatrix} \quad (4.16)$$

Whereas, in the CTRA model it is,

$$\begin{pmatrix} x_{t+1} \\ y_{t+1} \\ v_{t+1} \\ \theta_{t+1} \\ \omega_{t+1} \\ a_{t+1} \end{pmatrix} = \begin{pmatrix} x_t \\ y_t \\ v_t \\ \theta_t \\ \omega_t \\ a_t \end{pmatrix} + \begin{pmatrix} \frac{1}{\omega_t^2}[(v_t\omega_t + \mathrm{a}\omega_t \mathrm{T})\sin(\omega_t T + \theta_t) + a_t\cos(\theta_t + \omega_t T) - v_t\omega_t\sin\theta_t - a_t\cos\theta_t] \\ \frac{1}{\omega_t^2}[(-v_t\omega - \mathrm{a}\omega_t \mathrm{T})\cos(\omega_t T + \theta_t) + a_t\sin(\theta_t + \omega_t T) + v_t\omega_t\cos\theta_t - a_t\sin\theta_t] \\ v_t\sin(\theta_t + \omega_t T - \beta_t) \\ \omega_t T \\ 0 \\ 0 \end{pmatrix} \quad (4.17)$$

Note that T in (4.16) and (4.17) is the time interval for the equation calculation and β changes continuously depending on the simulated road central line parameter. d_{co} is calculated by the point to the tangent line of the curve distance formula.

4.3 Filter Model Design

Filter model design is critical for the estimated positioning performance, which is the key part of lane level irregular driving detection system developed in this thesis. As discussed in Chap. 3, Extended Kalman Filter (EKF) and Particle Filtering (PF) are the most relevant and applicable filters for sensor integration, and the design of the filter models based on EKF and PF are discussed in the following sub-section. The designed EKF and PF filtering models for irregular driving detection are totally new.

4.3.1 Extended Kalman Filter Based Precise Positioning Model Design

For the EKF based precise positioning system developed in this thesis, the state vector containing six parameters, is:

$$x(t) = (x\, y\, v\, \beta\, \theta\, \omega)^T \tag{4.18}$$

The measurement vector, containing four parameters measured from the GPS and INS is:

$$z(t) = (x\, y\, v\, \omega)^T \tag{4.19}$$

where x, is the X-axis coordinate (in metres) of point o in the local UK National Grid coordinate system; y, is the Y-axis coordinate (in metres) of point o in the local UK National Grid coordinate system; v, is the velocity at the heading direction; β, is the angle between the lane segment and the local British National Grid coordinates; θ, is the heading angle of the vehicle; ω, is the vehicle yaw rate.

The relationship between measurement vector $z(t)$ and state vector $x(t)$ is:

$$z(t) = Hx(t) \tag{4.20}$$

The Jacobian of the measurement model H is:

$$H = \begin{pmatrix} 1 & 0 & 0 & 0 & 0 & 0 \\ 0 & 1 & 0 & 0 & 0 & 0 \\ 0 & 0 & 1 & 0 & 0 & 0 \\ 0 & 0 & 0 & 0 & 1 & 0 \end{pmatrix} \tag{4.21}$$

The covariance matrix of measurement noise is:

$$R = \begin{pmatrix} \sigma_x^2 & 0 & 0 & 0 \\ 0 & \sigma_y^2 & 0 & 0 \\ 0 & 0 & \sigma_v^2 & 0 \\ 0 & 0 & 0 & \sigma_\omega^2 \end{pmatrix} \tag{4.22}$$

The Jacobian of the measurement model with respect to measurement noise is:

$$\mathrm{V} = \begin{pmatrix} \frac{\partial x}{\partial \sigma_x} & 0 & 0 & 0 \\ 0 & \frac{\partial y}{\partial \sigma_y} & 0 & 0 \\ 0 & 0 & \frac{\partial v}{\partial \sigma_v} & 0 \\ 0 & 0 & 0 & \frac{\partial \omega}{\partial \sigma_\omega} \end{pmatrix} \tag{4.23}$$

The estimated error covariance P is used together with the Jacobian matrix H and measurement noise covariance R together with the Jacobian matrix V to calculate the Kalman gain (Mohinder and Angus 2008). Once the Kalman Gain, K, is calculated, the system brings in the measured data z to correct the predicted parameters and also the covariance error. After correcting the previously predicted values, the system is ready to predict the next position by using the state vector equations. The filter also estimates the error covariance of the estimated parameters by using the

Jacobian of the system model with respect to state A and the Jacobian of the system model with respect to process noise W together with the process noise Q, as follows.

$$A = \left[\frac{\partial y}{\partial x} f(x)\right] \tag{4.24}$$

$$W = \left[\frac{\partial y}{\partial w} f(x)\right] \tag{4.25}$$

Process noise covariance:

$$Q = \begin{pmatrix} w_x^2 & & & & & \\ & w_y^2 & & & & \\ & & w_{v_{cl}}^2 & & & \\ & & & w_\beta^2 & & \\ & & & & w_\theta^2 & \\ & & & & & w_\omega^2 \end{pmatrix} \tag{4.26}$$

For the four vehicle motion models (CV, CA, CTRV, CTRA), each one has its own EKF model. For every prediction, d_{co} is calculated to be the minimum distance between $[x, y]$ and the lane central line as discussed before.

4.3.2 *Particle Filter Based Precise Positioning Model Design*

The PF is a non-parametric implementation of a recursive Bayes filter. The probability density is approximated by a number N of weighted samples. The steps for the designed lane level precise positioning algorithm based on PF are as follows (Gustafsson et al. 2002):

(1) Initialization

The defined state vector is given by:

$$x(t) = (x\, y\, v\, \theta\, \omega\, a\, \beta\, d_{co})^T \tag{4.27}$$

The state vector (4.27) can be divided into two sub state vectors. The stated vector (4.28) is defined as the vehicle motion vector, which is for the particle filter cycle, and state vector (4.29) is defined as the lane related vector, which is the dependent vector of state vector (4.28) in the calculation.

$$p(t) = (x\,y\,v\,\theta\,\omega\,a)^T \tag{4.28}$$

$$q(t) = (\beta d_{co})^T \tag{4.29}$$

In the particle filter operation, the parameters change with time epochs and particles with each parameter within the state vector (4.27) expressed as:

$$X_t^i(t = 0\ldots n; i = 1\ldots n) \tag{4.30}$$

where X_t^i, is the parameters within the state vector (4.27) at the time epoch t with the particle number i.

The filter begins with the initialization of the particles x_0^i of the vehicle motion vector $p(t)$. To realize this, first the local coordinate sub-state variables x and y are randomly generated following a Gaussian distribution with the first accepted GNSS point as the mean value and a standard deviation value according to the GNSS a posteriori solution statistics. The initial heading velocity v_0^i is set as 0, since the initial position of the vehicle is assumed as static. Because it is assumed that no information on the initial heading is available, the values of θ are uniformly spread through the whole range of 2 π space and the initial ω_0^i is 0.

For the initialization of $q(t)$, in the straight road section, β is constant, while in the curved road section β_0^i depends on $[x_0^i, y_0^i]$. β_0^i is the corresponding angle within the local coordinates frame of the lane central line. d_{co0}^i is the minimum distance between $[x_0^i,\ y_0^i]$ and the lane central line.

(2) Filter Prediction

The prediction of the $p(t)$ is calculated as follows

$$F(X) = \begin{pmatrix} x_{(t+1)}^i \\ y_{(t+1)}^i \\ v_{(t+1)}^i \\ \theta_{(t+1)}^i \\ \omega_{(t+1)}^i \\ a_{(t+1)}^i \end{pmatrix} = \begin{pmatrix} x_t^i \\ y_t^i \\ v_t^i \\ \theta_t^i \\ \omega_t^i \\ a_t^i \end{pmatrix} + \begin{pmatrix} \Delta_x^i \\ \Delta_y^i \\ \Delta_v^i \\ \Delta_\theta^i \\ \Delta_\omega^i \\ \Delta_a^i \end{pmatrix} \tag{4.31}$$

$\Delta_x^i, \Delta_y^i, \Delta_v^i, \Delta_\theta^i, \Delta_\omega^i,\ \Delta_a^i$ will be different from the different vehicle motion models. Constant velocity (CV) and Constant Acceleration (CA) assumptions can be applied on straight lane motion, while Constant Turn Rate and Velocity (CTRV) Constant Turn Rate and Acceleration (CTRA) perform a reasonable approximation of motions by vehicles on lanes with curves (Tsogas et al. 2005).

From the geometric relationship of the lane segment, the prediction of $q(t)$ can be expressed as:

$$\beta_{t+1}^{i} \approx \beta_{t}^{i} \tag{4.32}$$

$$d_{cot+1}^{i} = d_{cot}^{i} + \sin(\beta_{t}^{i})\Delta_{x}^{i} - \cos(\beta_{t}^{i})\Delta_{y}^{i} \tag{4.33}$$

(3) Filter update

The prediction cycle is applied at every input sample. First, the judgement of d_{co}^{i} is made. The valid d_{co}^{i} should comply with the equation $\left|d_{co}^{i}\right| < 3HL$ where, $3HL$ is 3 times half of the lane width. If d_{co}^{i} is larger than the border, then the particle has measurement error (Toledo-Moreo et al. 2010).

If d_{co}^{i} is within this interval, the prediction parameters of t + 1 are calculated. These predictions will only be considered as valid, however, when the position predicted for a particle i is still within the bounds of the lane width. After every prediction phase, therefore, the condition given by the following equation must be verified as follows.

If $|d_{co(t+1)}^{i}| < 3HL$ is satisfied and the predicted $d_{co(t+1)}^{i}$ is accepted, then the other predicted parameters are accepted. If $|d_{co(t+1)}^{i}| < 3HL$ is not satisfied, the other predicted parameters are considered as invalid and the weighting of the particle is thus set as $w_{0}^{i} = 0$. GNSS validity is tested after every prediction cycle and used to adjust the predicted particles to output the particle filter estimates.

4. Normalization and resampling

After every update phase, the weights of the particles are modified, and the normalization and resample test phases of the PF relaunched.

4.4 Fuzzy Inference System for Irregular Driving Detection

To detect the different types of driving pattern, the fuzzy inference system (FIS) can map the input to output using fuzzy logic by means of combination rules (Aljaafreh et al. 2012). The reason for choosing FIS as the algorithm to distinguish different types of irregular driving is that it does not require large numbers of samples compared to other pattern recognition methods such as neural networks or explicit modelling.

The inputs for FIS, O-indicator and D-indicator, are calculated on the MAD of *omega* and d with a window size of 1 s. The total collected one set of O-indicator and D-indicator for FIS input includes the preceding 5 s of data. The reason for using the last 5 s of data as the time interval is that in a time duration of 5 s, there

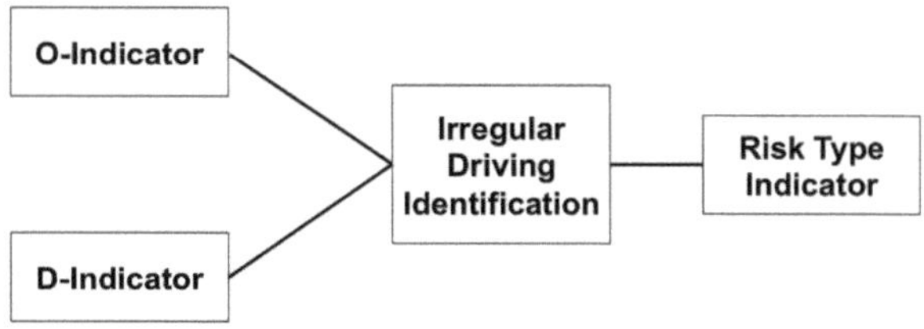

Fig. 4.5 FIS structure

are a total of 50 sets of data, which is sufficient to provide information for the FIS-based judgement of the driving style (Chang et al. 2008).

The O-indicator and D-indicator from the different integration models estimated results fusion model are the inputs of the FIS system to calculate their risk type indicators. Figure 4.5 shows the structure of the fuzzy inference system.

The designed membership function for the O-indicator and D-indicator values and risk type indicator for simulated straight and curved scenarios are shown in Figs. 4.6 and 4.7. The selection and formulation of the input and output fuzzy sets, and their membership function, are based on expert driver knowledge collected and classified from previous years' driver data (Aljaafreh et al. 2012). In Figs. 4.6 and 4.7, the first input is O-indicator data and corresponding fuzzy values in FIS are defined to be Small O-indicator (SO), Medium O-indicator (MO), Large O-indicator (LO) and Very Large O-indicator (VLO). The membership functions for the SO and VLO fuzzy sets are a trapezoidal function, while a triangular function is used for MO and LO. The second input is the D-indicator and the corresponding fuzzy values are defined to be Small D-indicator (SD), Medium D-indicator (MD) and Large D-indicator (LD) Very Large D-indicator (VLD). Finally, the output of the system is the risk type indicator, which is defined by four fuzzy values A, B, C, D. Fuzzy value A means low risk, B means medium risk, C means high risk and D means very high risk. The membership functions for A and D fuzzy sets are a trapezoidal function, while a triangular function is used for B and C.

Tables 4.1 and 4.2 show the rules for the defined FIS system for straight and curved lanes, based on the experience data. Figure 4.8 shows the designed FIS results in surface view, which presents a single pair of O-indicator value and D-indicator value, with one corresponding risk type indicator value.

Simulink is then used to output the risk type indicator values for every time epoch for both straight and curved scenarios, see Fig. 4.9. The driving classification indicator values are then developed and by comparing the sorting extracted from the driving classification indicators, the different driving styles can be distinguished. The details of how the driving classification indicator was developed and how the rules were extracted for the detection of irregular driving will be further discussed and analysed with a simulation in Chap. 5.

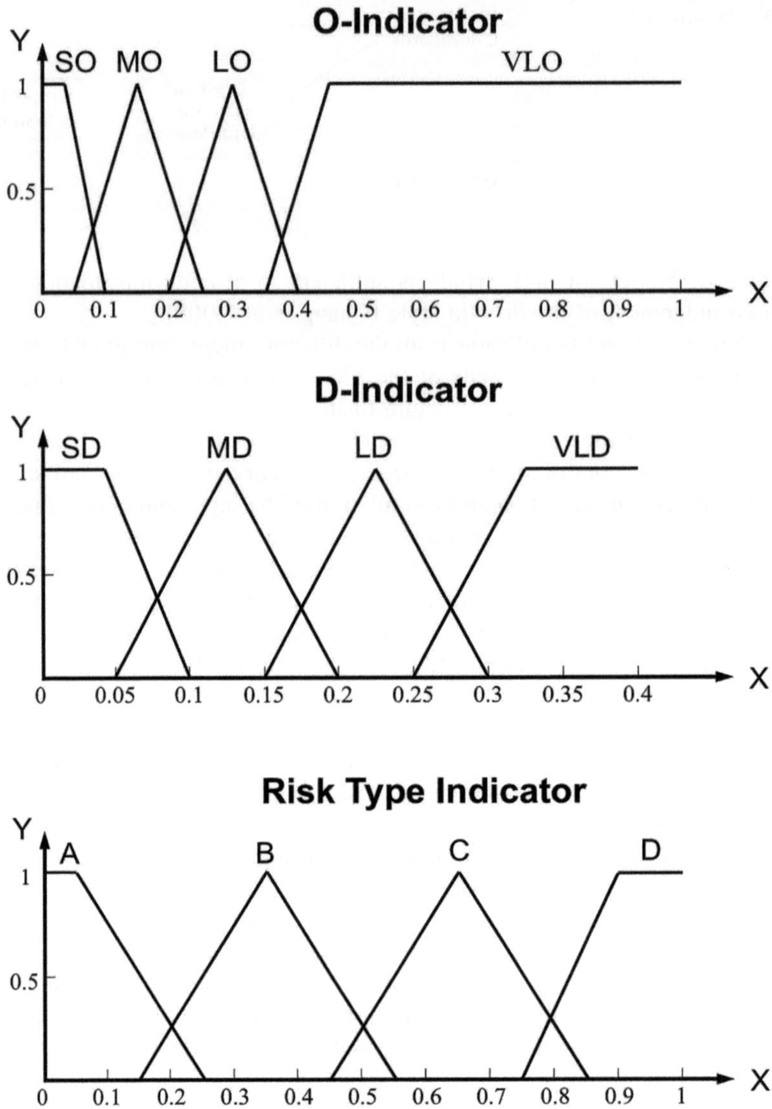

Fig. 4.6 Membership function for straight scenarios

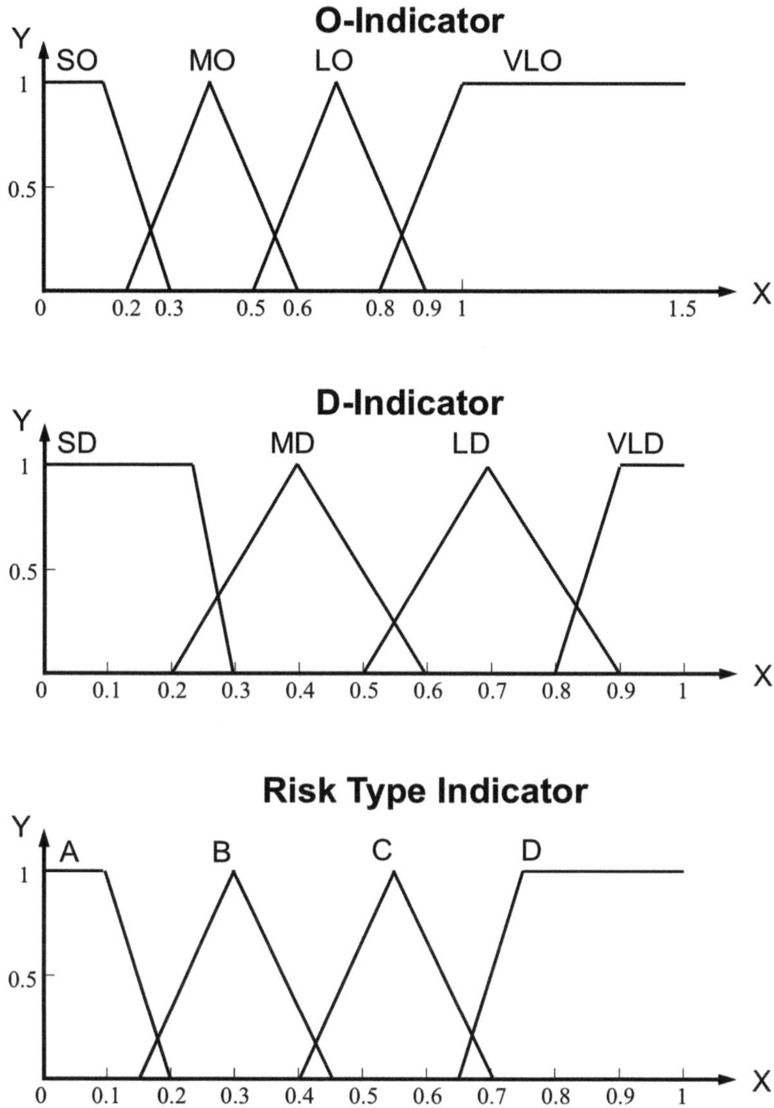

Fig. 4.7 Membership function for curved scenarios

Table 4.1 Rules for judgement of risk types on a straight lane

O-indicator	D-indicator	Risk type indicator
SO	SD	A
	MD	B
	LD	C
	VLD	
MO	Any (SD, MD, LD, VLD)	C
LO	Any (SD, MD, LD, VLD)	D
VLO	Any (SD, MD, LD, VLD)	D

Table 4.2 Rules for judgement of risk types on a curved lane

O-indicator	D-indicator	Risk type indicator
SO	SD	A
	MD	B
	LD	D
	VLD	
MO	SD	B
	MD	
	LD	D
	VLD	
LO	SD	C
	MD	
	LD	
	VLD	D
VLO	SD	C
	MD	
	LD	
	VLD	D

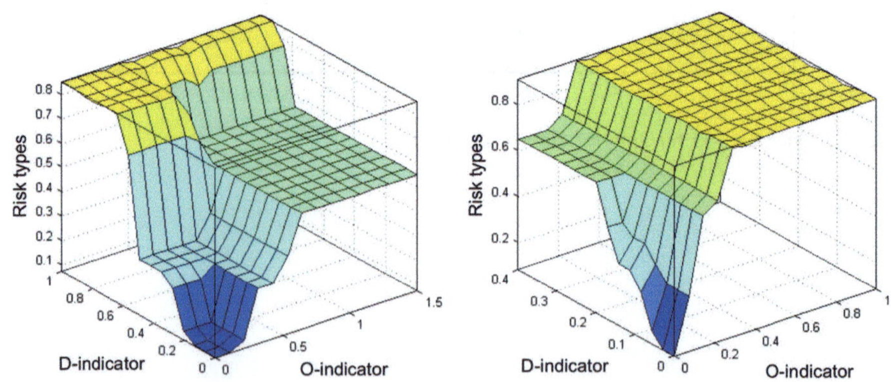

Fig. 4.8 Designed FIS surface in logic view

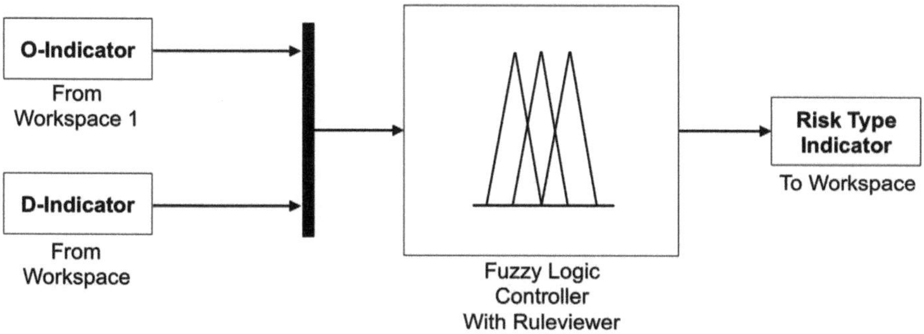

Fig. 4.9 Simulink simulation structure

4.5 Summary

In this chapter, first, the overview and framework of a lane level irregular driving detection algorithm has been introduced. Then, the precise vehicle kinematic model design was presented based on the road segment model and precise vehicle motion model on straight and curved road sections. This was followed by the design of the filter models, which are mainly based on the PF and EKF. Finally, the FIS membership functions and rules for the irregular driving detection are designed. In the next chapter, the algorithms presented in this chapter are implemented based on the simulation and field test.

References

Aljaafreh, A., Alshabatat, N., & Najim Al-Din, M. S. (2012). Driving style recognition using fuzzy logic. *IEEE International Conference on Vehicular Electronics and Safety (ICVES), 2012*, 460–463.

Chang, T. H., Hsu, C. S., Wang, C., & Yang, L. K. (2008). Onboard measurement and warning module for irregular vehicle behaviour. *IEEE Transactions on Intelligent Transportation Systems, 9*(3), 501–513.

Gustafsson, F., Gunnarsson, F., Bergman, N., Forssell, U., Jansson, J., Karlsson, R., et al. (2002). Particle filters for positioning, navigation and tracking. *IEEE Transactions on Signal Processing, 50*, 425–435.

Mohinder, S. G., & Angus, P. A. (2008). *Kalman filtering: Theory and practice using MATLAB*. New York: Wiley-IEEE Press. ISBN: 978-0-470-17366-4.

NHTSA. (2010). http://www.nhtsa.gov

Ren, X., & Cai, Z. (2010). Kinematics model of unmanned driving vehicle. In *Proceedings of the 8th world congress on intelligent control and automation, Jinan, China.*

Schubert, R., Richter, E., & Wanielik, G. (2008). Comparison and evaluation of advanced motion models for vehicle tracking. In *2008 11th international conference on information fusion* (pp. 1–6).

Toledo-Moreo, R., Bétaille, D., & Peyret, F. (2010). Lane-level integrity provision for navigation and map matching with GNSS, dead reckoning, and enhanced maps. *IEEE Transactions on Intelligent Transportation Systems, 11*(1).

Toledo-Moreo, R., Bétaille, D., Peyret, F., & Laneurit, J. (2009). Fusing GNSS, dead-reckoning, and enhanced maps for road vehicle lane-level navigation. *IEEE Journal of Selected Topics in Signal Processing, 3*(5).

Tsogas, M., Polychronopoulos, A., & Amditis, A. (2005). Unscented Kalman filter design for curvilinear motion models suitable for automotive safety applications. In *Proceedings of the IEEE 7th international conference on information fusion* (pp. 1295–1302).

Chapter 5
Testing, Analysis and Performance Validation

Abstract In this chapter, the defined reference data for weaving, swerving, jerky driving and normal driving, both in straight and curved lanes scenarios are created in the simulation, and the O-indicator and D-indicator from the reference are calculated as the input of the FIS system to calculate the reference risk type indicator values and driving classification indicator values. The sorting rules for the driving classification indicators from the reference are extracted to arrive at the driving style judgement. The simulated GPS and INS data are then generated with the designed filter and motion model combination to calculate the precise positioning and dynamic parameter results used for the different types of driving style identification. By comparing the sorting rules of the driving classification indicator calculated from the filter estimated results and reference results, the different types of driving styles based on the filter estimated results can be detected. Finally, real field data based tests are carried out to validate the simulation results. Because of logistical and safety concerns in respect to undertaking field tests, the approach taken in this chapter is to address simulation in detail with limited but targeted field tests for the overall validation of the simulation results.

In the previous chapters, a lane level irregular driving identification system has been designed on an integrated positioning system for precise positioning and dynamic parameters estimation, together with an FIS based algorithm for irregular driving detection. In this chapter, the defined reference data for weaving, swerving, jerky driving and normal driving, both in straight and curved lanes scenarios are created in the simulation, and the O-indicator and D-indicator from the reference are calculated as the input of the FIS system to calculate the reference risk type indicator values and driving classification indicator values. The sorting rules for the driving classification indicators from the reference are extracted to arrive at the driving style judgement. The simulated GPS and INS data are then generated with the designed filter and motion model combination to calculate the precise positioning and dynamic parameter results used for the different types of driving style identification. By comparing the sorting rules of the driving classification indicator calculated from the filter estimated results and reference results, the different types of driving

R. Sun, *An Integrated Solution Based Irregular Driving Detection*,
Springer Theses, DOI 10.1007/978-3-319-44926-5_5

styles based on the filter estimated results can be detected. Finally, real field data based tests are carried out to validate the simulation results. Because of logistical and safety concerns in respect to undertaking field tests, the approach taken in this chapter is to address simulation in detail with limited but targeted field tests for the overall validation of the simulation results.

5.1 Definition of Scenarios and Rules Extraction for Irregular Driving Detection

The scenarios are generated based on the extracted true driving styles on straight roads and curved roads. The designed lane width is set as 3.5 m since the regularised lane width of both motorway and city roads in UK is 3.5 m, and the simulated vehicle is defined to be 1 m wide for easy calculation, the relationship between the vehicle width and its effect on irregular driving detection will be discussed later. If the vehicle central line and the road central line overlap, therefore, the lateral gap from the vehicle edge to the lane marking is 1.25 m. Thus, if the vehicle veers towards the right lane border, it crosses the lane marking if the lateral displacement approaches 1.25 m, which means the maximum lateral displacement is 1.25 m for the vehicle within the lane. A simple vehicle kinematic model is used to generate the reference trajectory of the vehicle with the heading velocity of 85–120 km/h. The model has taken into account the vehicle's velocity, heading and yaw rate. Using the yaw rate and lateral displacement as input, the vehicle's easting and northing coordinates can be determined, see (5.1), (5.2) and (5.3) (Clanton et al. 2009).

The simple equations are

$$\dot{X} = v\sin(\theta) \tag{5.1}$$

$$\dot{Y} = v\cos(\theta) \tag{5.2}$$

$$\theta = \frac{v}{l}\tan(\omega t) \tag{5.3}$$

where X is the easting of the local coordinates of the vehicle; Y is the northing of the local coordinates of the vehicle; v is the velocity of the vehicle; θ is the heading angle of the vehicle; ω is the yaw rate; l is the vehicle wheelbase.

In the straight scenarios, the simulated reference trajectories are designed to drive along the X-axis to make the calculation simpler. Thus, only northing coordinates are considered, which means only lateral position is considered.

For the generation of the reference trajectory of Scenario 1, the kinematic motion model in (5.1), (5.2) and (5.3) is used. In addition, an amplitude sinusoidal yaw rate is used to generate a slow oscillation within the lane, because it follows the features of weaving. Figure 5.1 shows the weaving trajectory generated. Generation of

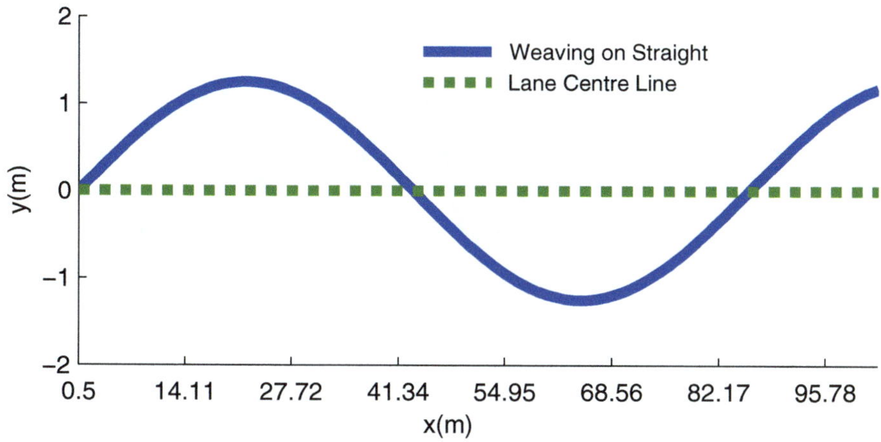

Fig. 5.1 S1 weaving on straight

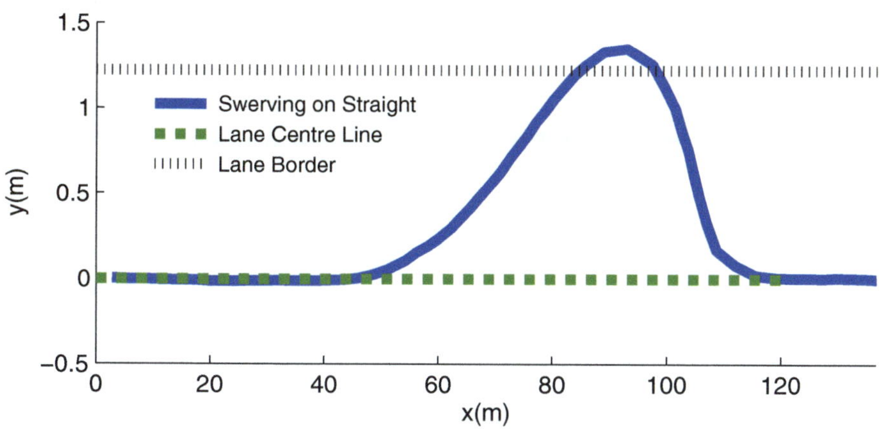

Fig. 5.2 S2 swerving on straight

Scenario 2 is adding a suddenly big yaw rate and then decreasing the lateral displacement gradually and back again sharply, see Fig. 5.2. Scenario 3 is generated by a very quick change in the yaw rate and lateral displacement input of the kinematic motion model with a higher frequency compared to Scenario 1, see Fig. 5.3. Scenario 4 is generated with a yaw rate and lateral displacement with very small noise added, see Fig. 5.4.

The curved scenarios are generated based on the curved lane created by PTV VISSIM software (PTV VISSIM 2013). The curved lane central line is generated by the simulated vehicle driving along the middle of the lane and outputting the points of the trajectory coordinates with 10 Hz. Scenario 5 is generated with the sinusoidal amplitude added following the curved lane, see Fig. 5.5. Scenario 6 is

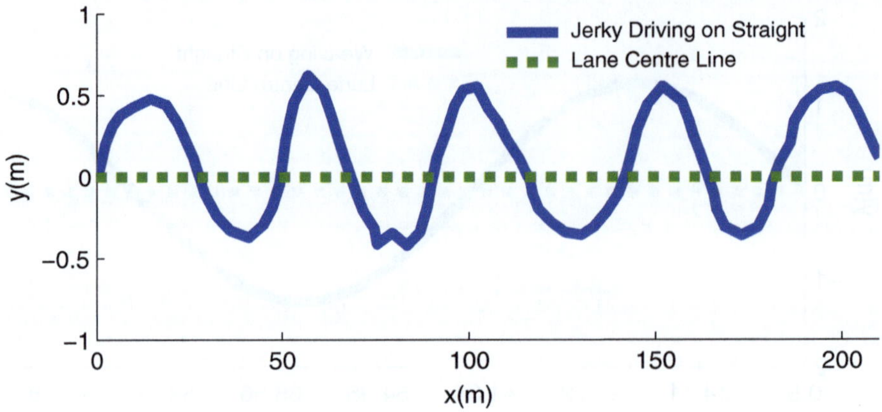

Fig. 5.3 S3 jerky driving on straight

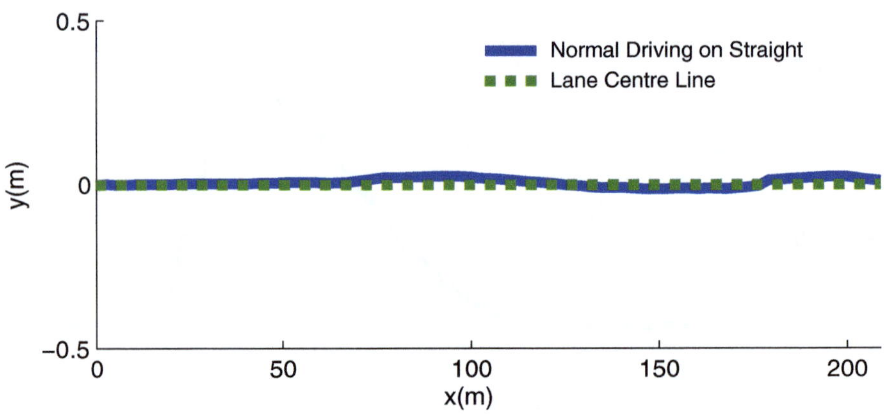

Fig. 5.4 S4 normal driving on straight

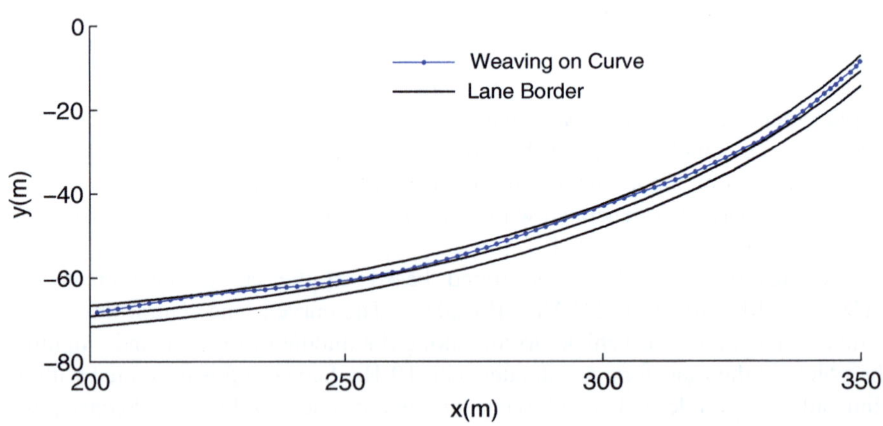

Fig. 5.5 S5 weaving on curve

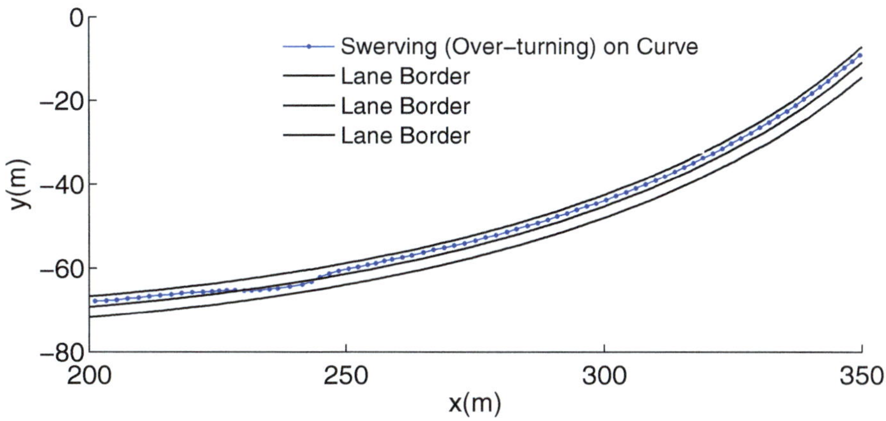

Fig. 5.6 S6a swerving (over-turning) on curve

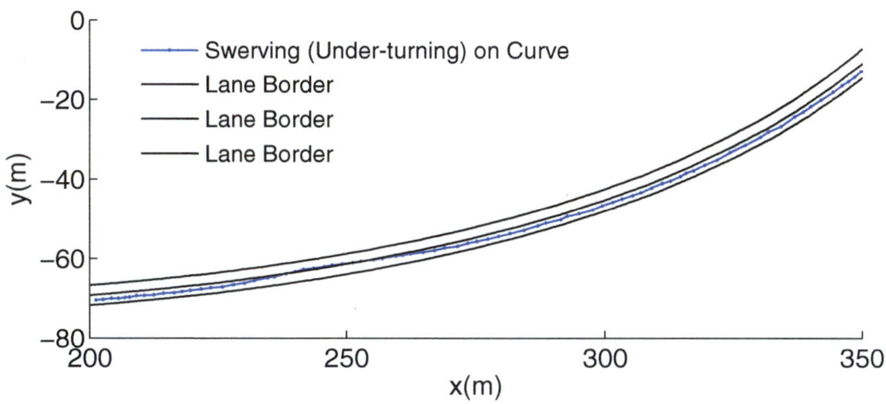

Fig. 5.7 S6b swerving (under-turning) on curve

defined as swerving on a curved lane. Swerving on a curve includes over-turning and under-turning, although these are treated as the same scenario because they are similar manoeuvres. Over-turning and under-turning are generated by suddenly big or small yaw rates and then keep a similar heading angle with the lane curvature, see Figs. 5.6 and 5.7. Scenario 7, jerky driving on a curve, has been generated by a quick change in yaw rate and lateral displacement input of the kinematic motion model and with a constant yaw rate to follow the curved lane, see Fig. 5.8. Scenario 8, normal driving on a curve, is generated by a constant small yaw rate and very small lateral displacement with small noise added, see Fig. 5.9.

For reference data in both straight lane and curved lane scenarios, in order to distinguish the features of the driving styles, as discussed in Chap. 4, the Moving Average Deviation (MAD) with a window size of 10 data points (i.e. 1 s) is applied

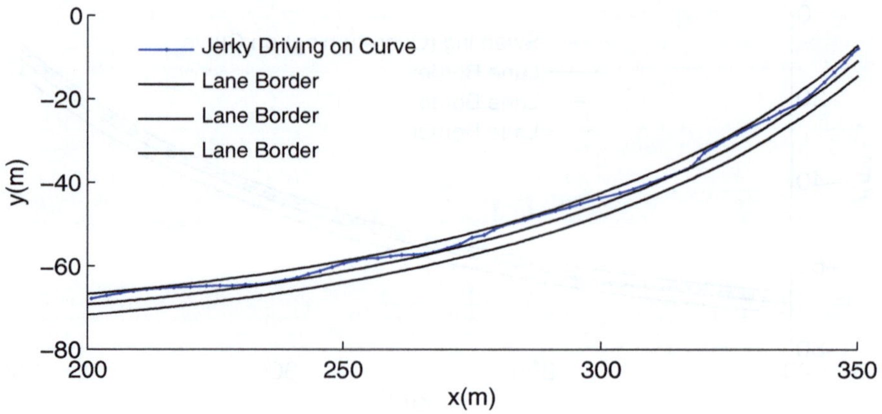

Fig. 5.8 S7 jerky driving on curve

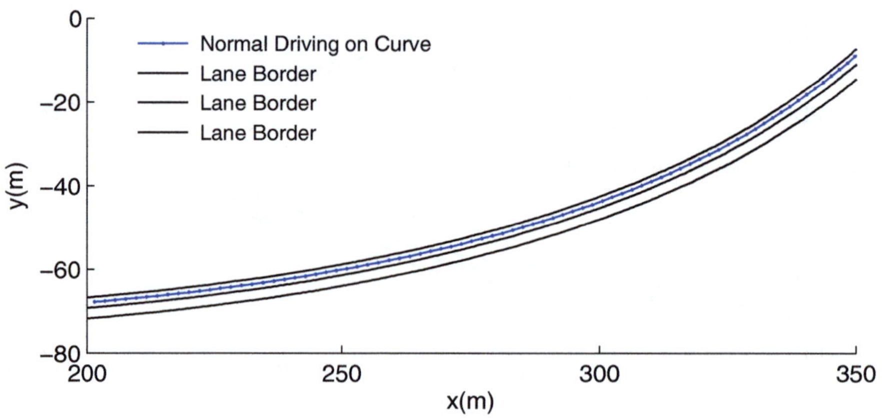

Fig. 5.9 S8 normal driving on curve

on the *omega* and *d* to calculate the O-indicator and D-indicator. The average deviation is that of the absolute deviations of the data points from their mean, which is always used to measure the variability of a data set. The comparison of the O-indicator and D-indicator for the scenarios S1, S2, S3, and S4 is shown in Figs. 5.10 and 5.11. It is indicated that for different scenarios, their O-indicator or D-indicator has a different range of values. For example, most of the O-indicator values for jerky driving cover the range between 0.15 and 0.6 rad, which is much higher than the value of weaving and swerving (below 0.15 rad). The comparison of the O-indicator and D-indicator values for the scenarios S5, S6a, S6b, S7, and S8 is shown in Figs. 5.12 and 5.13. The performance is similar to the corresponding straight scenarios, which also implies that the range of the values of the O-indicator and D-indicator for different scenarios is different. This, therefore, is the principle for distinguishing between the different types of irregular driving styles.

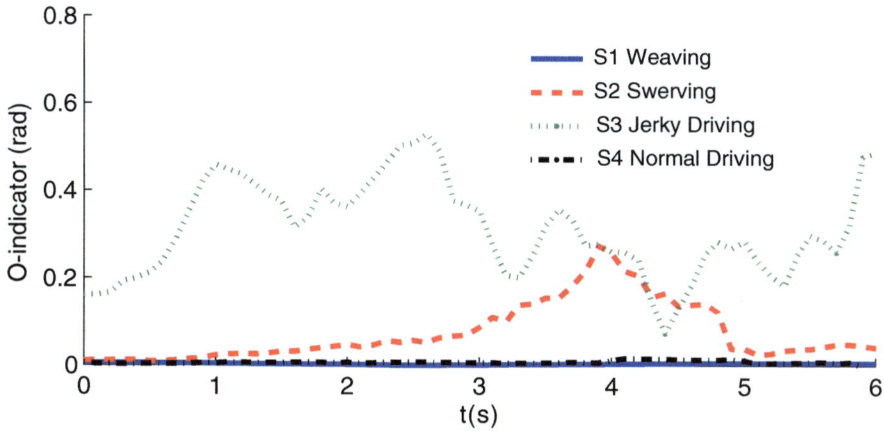

Fig. 5.10 S1 S2 S3 S4 straight lane scenarios O-indicator comparison based on reference data

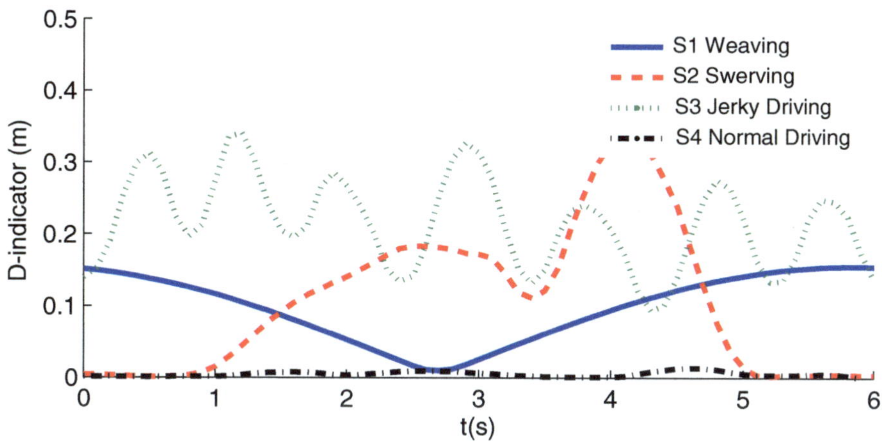

Fig. 5.11 S1 S2 S3 S4 straight lane scenarios D-indicator comparison based on reference data

By applying the FIS algorithm designed in 4.4, the output risk type indicator will be discussed as follows. The 5 s of risk type indicator output results from FIS for straight scenarios based on the simulated 'reference data' are plotted in Fig. 5.14. Table 5.1 shows the statistics of the number of points in each risk type for the straight scenarios.

Similarly, for the curved scenarios, the risk type indicator results are plotted in Fig. 5.15. The number of points of the risk type indicator in each risk type are shown in Table 5.2.

From the analyses in Tables 5.1 and 5.2, it can be seen that each scenario returns one or two dominant risk types in the 5 s sample. If there was only one dominant risk type for a given scenario, containing the most points, it would be straightforward to

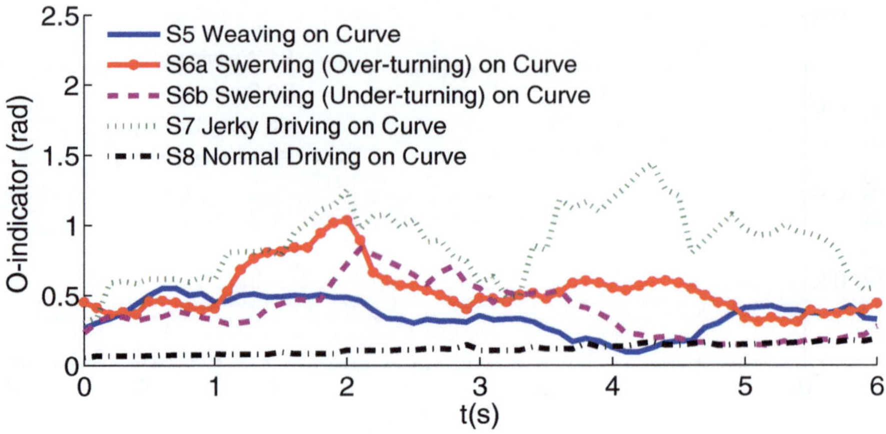

Fig. 5.12 S5 S6a S6b S7 S8 curved lane scenarios O-indicator comparison based on reference data

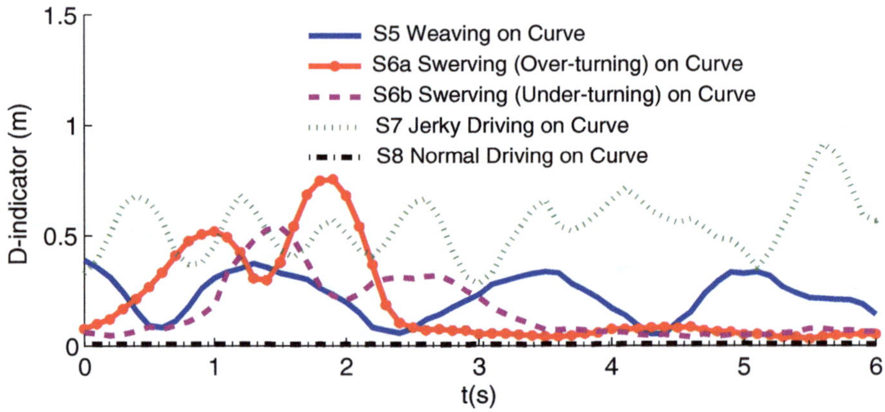

Fig. 5.13 S5 S6a S6b S7 S8 curved lane scenarios D-indicator comparison based on reference data

classify the scenarios. This is not always the case, however, with two or more risk types having a similar number of points for some scenarios. For example, in S7, there are two dominant risk types with 25 in risk type D and 22 in risk type C. In order to amplify the features of each scenario, the points in relevant risk types are combined as shown in Tables 5.3 and 5.4, effectively resulting in a new indicator for the classification of driving. The new indicator is named a driving classification indicator. The four parameters in each driving classification indicator are developed based on the sum number of points in relevant risk types, e.g., AB is the sum number

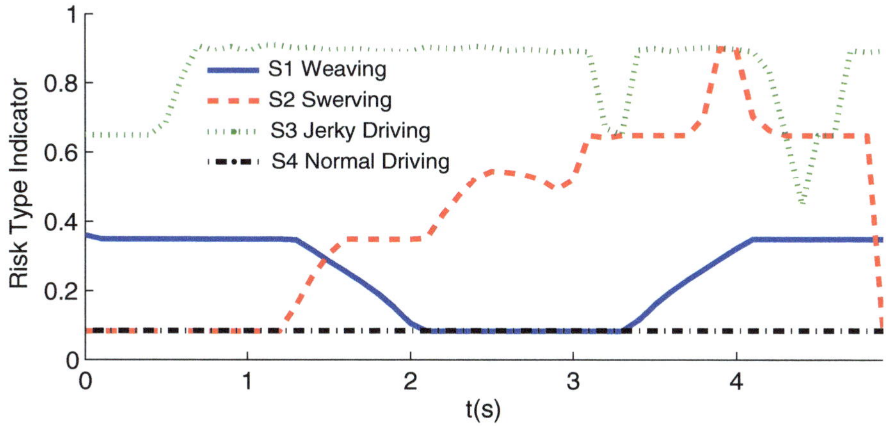

Fig. 5.14 Risk type indicator comparison for straight scenarios based on reference data

Table 5.1 Statistics of number of points in risk types for straight lane scenarios based on reference data

Number of points	Risk types			
Scenarios	A	B	C	D
S1 weaving	19	31	0	0
S2 swerving	15	11	22	2
S3 jerky driving	0	1	11	38
S4 normal driving	50	0	0	0

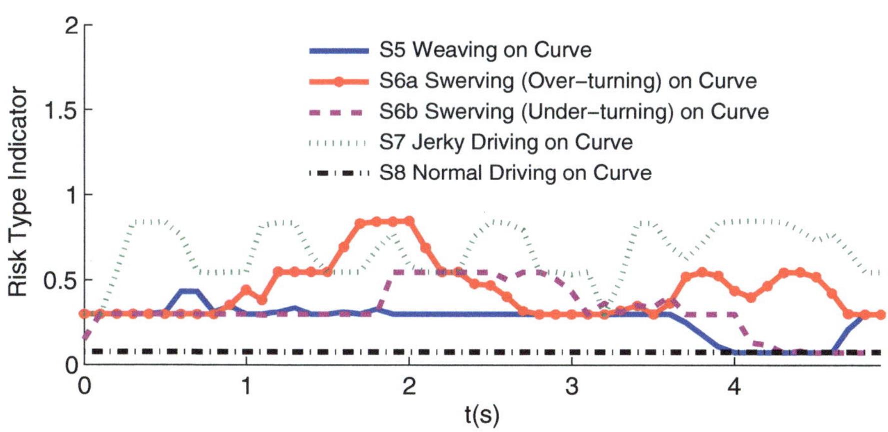

Fig. 5.15 Risk type indicator comparison for curved scenarios based on reference data

Table 5.2 Statistics of number of points in risk types for curved scenarios based on reference data

Number of points	Risk types			
Scenarios	A	B	C	D
S5 weaving on curve	8	40	2	0
S6a swerving (over-turning) on curve	0	25	19	6
S6b swerving (under-turning) on curve	10	28	12	0
S7 jerky driving on curve	0	3	22	25
S8 normal driving on curve	50	0	0	0

Table 5.3 Driving classification indicators for straight scenarios based on reference

Scenarios	AB	BC	CD	AD
S1 weaving	50	31	0	19
S2 swerving	26	33	24	17
S3 jerky driving	1	12	49	38
S4 normal driving	50	0	0	0

Table 5.4 Driving classification indicators for curved scenarios based on reference

Scenarios	AB	BC	CD	AD
S5 weaving on curve	48	42	2	8
S6a swerving (over-turning) on curve	25	44	25	6
S6b swerving (under-turning) on curve	38	40	12	10
S7 jerky driving on curve	3	25	47	25
S8 normal driving on curve	50	0	0	50

of risk type A and risk type B; BC is the sum number of risk type B and risk type C; CD is the sum number of risk type C and risk type D; AD is the sum number of risk type A and D.

Tables 5.3 and 5.4 show the calculated driving classification indicator results for all straight and curved scenarios. The parameters of each driving style classification indicator are calculated and compared to determine the feature of each scenario. Based on the calculation and comparison results of the AB, BC, CD and AD parameters, the sorting rules can be extracted from the driving classification indicator values based on the reference data for each scenario. For example, the rule AB > BC > CD > AD is extracted for weaving from S1, while for S7, CD > BC = AD > AB is extracted as jerky driving. Thus, based on analysis of all scenarios, the rules for detecting the features of each scenario are extracted in Table 5.5.

These rules extracted from the reference data for the judgement of different types of irregular driving styles are applied in Sect. 5.2 for sensitivity analysis.

Table 5.5 Sorting rules for driving style judgement

No.	Sorting rules	Judgement
1	AB > BC >= CD >= AD	Weaving/weaving on curve
2	AB > BC >= AD >= CD	Weaving/weaving on curve
3	AB > CD >= BC >= AD	N/A
4	AB > CD >= AD >= BC	Weaving/weaving on curve
5	AB > AD >= BC >= CD	Normal driving/normal driving on curve
6	AB > AD >= CD >= BC	N/A
7	BC > CD >= AD >= AB	Swerving/over-turning (under-turning)
8	BC > CD >= AB >= AD	Swerving/over-turning (under-turning)
9	BC > AD >= AB >= CD	N/A
10	BC > AD >= CD >= AB	N/A
11	BC > AB >= CD >= AD	Swerving/over-turning (under-turning)
12	BC > AB >= AD >= CD	Swerving/over-turning (under-turning)
13	CD > AD >= AB >= BC	Jerky driving/jerky driving on curve
14	CD > AD >= BC >= AB	Jerky driving/jerky driving on curve
15	CD > AB >= BC >= AD	N/A
16	CD > AB >= AD >= BC	N/A
17	CD > BC >= AB >= AD	N/A
18	CD > BC >= AD >= AB	Jerky driving/jerky driving on curve
19	AD >= AB >= BC >= CD	Normal driving/normal driving on curve
20	AD > AB >= CD >= BC	Normal driving/normal driving on curve
21	AD > CD >= AB >= BC	Jerky driving/jerky driving on curve
22	AD > CD >= BC >= AB	Jerky driving/jerky driving on curve
23	AD > BC >= AB >= AD	N/A
24	AD > BC >= AD >= AB	N/A
25	AB = AD	Normal driving/normal driving on curve

5.2 Sensitivity Analysis of Irregular Driving Detection Algorithms

The sensitivity analysis is designed to test how well the designed irregular driving detection algorithm performs in relation to its input parameters. In order to make the simulation tracks simpler and more comparable, sinusoidal waves with different periods and amplitudes were generated as the input for the test. If the extracted irregular driving detection rules are applied to the designed scenarios, it is evident that the irregular driving detection algorithm returns the same characteristics for both the straight and curved scenarios. The sensitivity analysis based on straight scenarios can therefore represent both straight and curved scenarios.

These tracks are not marked as any defined driving styles and are just to see how the designed algorithms output to the input driving tracks. The reason for using

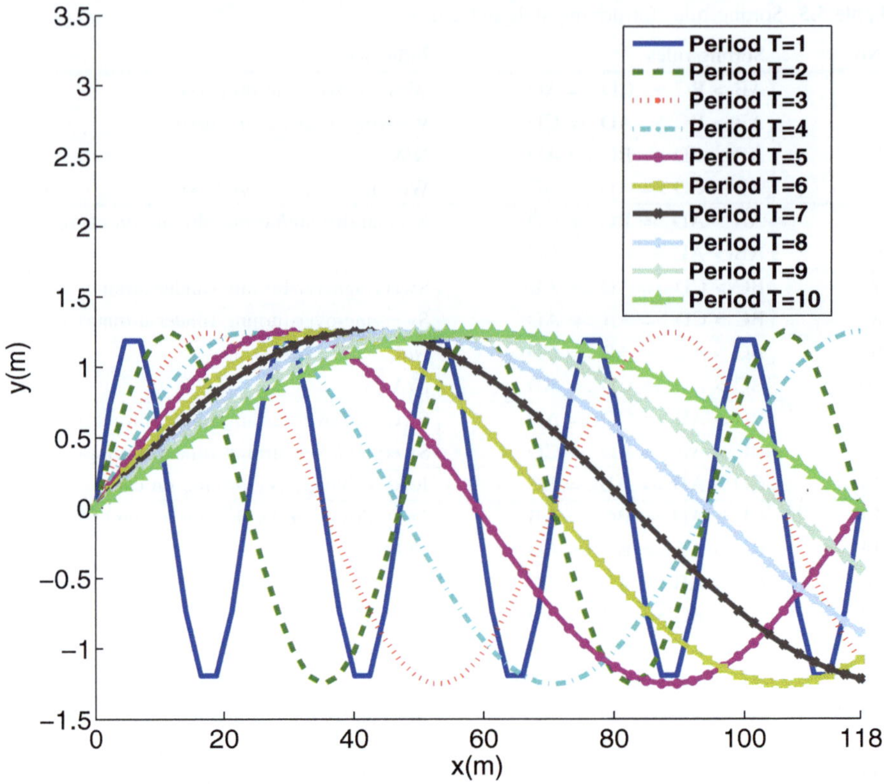

Fig. 5.16 Simulated vehicle track input for straight scenarios with various periods and fixed amplitude of 1.25 m

sinusoidal waves is that generating the waves is easy and changes in the amplitudes and periods of the waves can represent different driving styles.

The simulated input tracks were based on a velocity on the x-axis of about 85 km/h. Based on this velocity, the 70 sets of tracks with different periods and amplitudes representing unknown driving styles are input for processing through the irregular detection algorithm. The input datasets covered the period from 1 to 10 s, incrementing at 1 s intervals. For every fixed input period, the amplitude changes from 0.05 to 1.75 m (the possible range for the vehicle's lateral displacement), incrementing at 0.25 m intervals. Figure 5.16 is an example of the simulated vehicle track input at the velocity of 85 km/h with changing periods from 1 to 10 s and a fixed amplitude of 1.25 m. It is shown that within 5 s, in the fixed amplitude and increasing period, the tracks become flatter and smoother. Figure 5.17 is an example for a simulated vehicle track where the amplitude changes from 0.25 to 1.75 m but the period is fixed (T = 5). Here, as the amplitude becomes smaller, the tracks become flatter and smoother. Tables 5.6, 5.7, 5.8, 5.9, 5.10, 5.11, 5.12, 5.13, 5.14 and 5.15 show the calculated driving classification

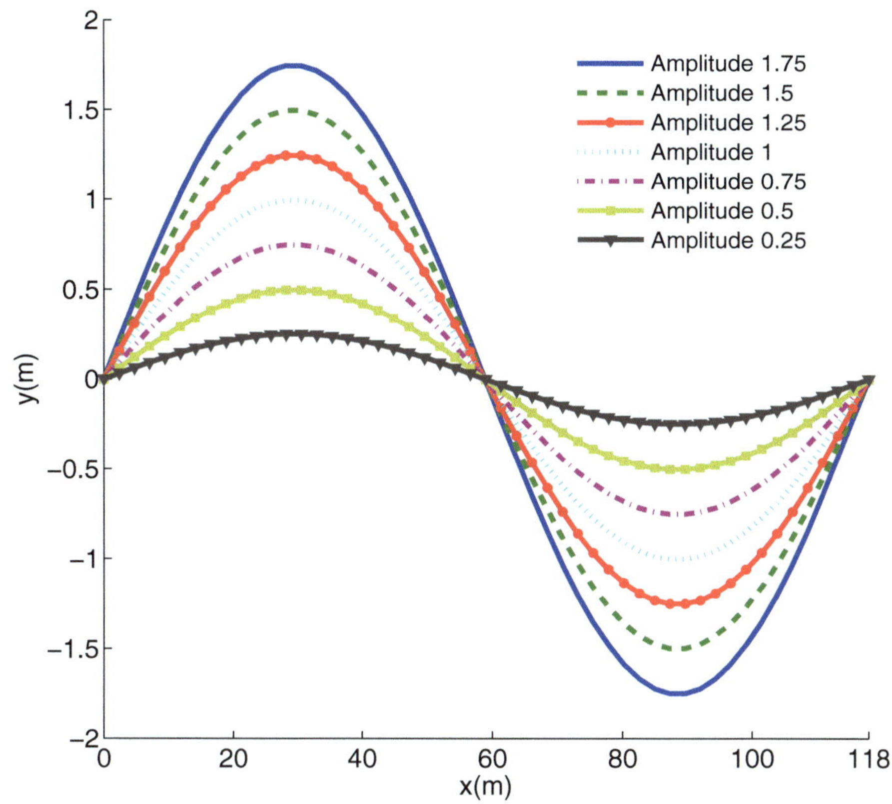

Fig. 5.17 Simulated vehicle track input for straight scenarios with fixed period (T = 5) and various amplitudes

Table 5.6 The driving classification indicator and driving style judgement results for input sinusoidal track with T = 1

Amplitude (m) (T = 1)	AB	BC	CD	AD	Results	Judgement
1.75	0	0	50	50	AD = CD > AB = BC	Jerky driving
1.5	0	0	50	50	AD = CD > AB = BC	Jerky driving
1.25	0	0	50	50	AD = CD > AB = BC	Jerky driving
1	0	0	50	50	AD = CD > AB = BC	Jerky driving
0.75	0	0	50	50	AD = CD > AB = BC	Jerky driving
0.5	0	0	50	50	AD = CD > AB = BC	Jerky driving
0.25	0	0	50	50	AD = CD > AB = BC	Jerky driving

Table 5.7 The driving classification indicator and driving style judgement results for input sinusoidal track with T = 2

Amplitude (m) (T = 2)	AB	BC	CD	AD	Results	Judgement
1.75	0	0	50	50	AD = CD > BC = AB	Jerky driving
1.5	0	0	50	50	AD = CD > BC = AB	Jerky driving
1.25	0	10	50	40	CD > AD > BC > AB	Jerky driving
1	0	20	50	30	CD > AD > BC > AB	Jerky driving
0.75	0	30	50	20	CD > BC > AD > AB	Jerky driving
0.5	10	50	40	0	BC > CD > AB > AD	Swerving
0.25	30	25	20	25	AB > BC = AD > CD	Weaving

Table 5.8 The driving classification indicator and driving style judgement results for input sinusoidal track with T = 3

Amplitude (m) (T = 3)	AB	BC	CD	AD	Results	Judgement
1.75	0	26	50	24	CD > BC > AD > AB	Jerky driving
1.5	0	26	50	24	CD > BC > AD > AB	Jerky driving
1.25	4	38	46	12	CD > BC > AD > AB	Jerky driving
1	12	50	38	0	BC > CD > AB > AD	Swerving
0.75	19	50	31	0	BC > CD > AB > AD	Swerving
0.5	26	46	24	4	BC > AB > CD > AD	Swerving
0.25	50	31	0	19	AB > BC > AD > CD	Weaving

Table 5.9 The driving classification indicator and driving style judgement results for input sinusoidal track with T = 4

Amplitude (m) (T = 4)	AB	BC	CD	AD	Results	Judgement
1.75	6	34	44	16	CD > BC > AD > AB	Jerky driving
1.5	12	42	38	8	BC > CD > AB > AD	Swerving
1.25	12	50	38	0	BC > CD > AB > AD	Swerving
1	18	50	32	0	BC > CD > AB > AD	Swerving
0.75	24	44	26	6	BC > CD > AB > AD	Swerving
0.5	42	38	8	12	AB > BC > AD > CD	Weaving
0.25	50	21	0	29	AB > AD > BC > CD	Normal driving

indicator values based on the output of the designed algorithms and the corresponding sorting of the indicators, and detected driving styles for the input track set with specific defined periods.

Based on the analysis from Tables 5.6, 5.7, 5.8, 5.9, 5.10, 5.11, 5.12, 5.13, 5.14, 5.15 and 5.16 is created to show the overall results of the input driving tracks: the 'blue blocks' in the table represent the detected jerky driving output, which only exist when the period of the driving track is less than 5 s. The 'green blocks' show

Table 5.10 The driving classification indicator and driving style judgement results for input sinusoidal track with T = 5

Amplitude (m) (T = 5)	AB	BC	CD	AD	Results	Judgement
1.75	14	50	36	0	BC > CD > AB > AD	Swerving
1.5	10	50	40	0	BC > CD > AB > AD	Swerving
1.25	14	48	36	2	BC > CD > AB > AD	Swerving
1	18	44	32	6	BC > CD > AB > AD	Swerving
0.75	26	40	24	10	BC > CD > AB > AD	Swerving
0.5	50	36	0	14	AB > BC > AD > CD	Weaving
0.25	50	16	0	34	AB > AD > BC > CD	Normal driving

Table 5.11 The driving classification indicator and driving style judgement results for input sinusoidal track with T = 6

Amplitude (m) (T = 6)	AB	BC	CD	AD	Results	Judgement
1.75	16	46	34	4	BC > CD > AB > AD	Swerving
1.5	16	46	34	4	BC > CD > AB > AD	Swerving
1.25	20	42	30	8	BC > CD > AB > AD	Swerving
1	28	42	22	8	BC > AB > CD > AD	Swerving
0.75	42	38	8	12	AB > BC > AD > CD	Weaving
0.5	50	30	0	20	AB > BC > AD > CD	Weaving
0.25	50	0	0	50	AB = AD	Normal driving

Table 5.12 The driving classification indicator and driving style judgement results for input sinusoidal track with T = 7

Amplitude (m) (T = 7)	AB	BC	CD	AD	Results	Judgement
1.75	18	44	32	6	BC > CD > AB > AD	Swerving
1.5	21	41	29	9	BC > CD > AB > AD	Swerving
1.25	24	41	26	9	BC > CD > AB > AD	Swerving
1	33	38	17	12	BC > AB > CD > AD	Swerving
0.75	50	35	0	15	AB > BC > AD > CD	Weaving
0.5	50	26	0	24	AB > BC > AD > CD	Weaving
0.25	50	0	0	50	AB = AD	Normal driving

Table 5.13 The driving classification indicator and driving style judgement results for input sinusoidal track with T = 8

Amplitude (m) (T = 8)	AB	BC	CD	AD	Results	Judgement
1.75	15	46	35	4	BC > CD > AB > AD	Swerving
1.5	5	50	45	0	BC > CD > AB > AD	Swerving
1.25	24	44	26	6	BC > CD > AB > AD	Swerving
1	36	42	14	8	BC > AB > CD > AD	Swerving
0.75	50	38	0	12	AB > BC > AD > CD	Weaving
0.5	50	26	0	24	AB > BC > AD > CD	Weaving
0.25	50	0	0	50	AB = AD	Normal driving

Table 5.14 The driving classification indicator and driving style judgement results for input sinusoidal track with T = 9

Amplitude (m) (T = 9)	AB	BC	CD	AD	Results	Judgement
1.75	17	43	33	7	BC > CD > AB > AD	Swerving
1.5	21	43	29	7	BC > CD > AB > AD	Swerving
1.25	27	41	23	9	BC > AB > CD > AD	Swerving
1	50	39	0	11	AB > BC > AD > CD	Weaving
0.75	50	35	0	15	AB > BC > AD > CD	Weaving
0.5	50	23	0	27	AB > AD > BC > CD	Normal driving
0.25	50	0	0	50	AB = AD	Normal driving

Table 5.15 The driving classification indicator and driving style judgement results for input sinusoidal track with T = 10

Amplitude (m) (T = 10)	AB	BC	CD	AD	Results	Judgement
1.75	22	42	28	8	BC > CD > AB > AD	Swerving
1.5	26	40	24	10	BC > AB > CD > AD	Swerving
1.25	36	38	14	12	BC > AB > CD > AD	Swerving
1	50	36	0	14	AB > BC > AD > CD	Weaving
0.75	50	30	0	20	AB > BC > AD > CD	Weaving
0.5	50	16	0	34	AB > AD > BC > CD	Normal driving
0.25	50	0	0	50	AB = AD	Normal driving

Table 5.16 Table of judgement results for the input track with various periods and amplitudes

Period(s) / Amplitude(m)	1	2	3	4	5	6	7	8	9	10
1.75	Jerky Driving	Jerky Driving	Jerky Driving	Jerky Driving	Swerving	Swerving	Swerving	Swerving	Swerving	Swerving
1.5	Jerky Driving	Jerky Driving	Jerky Driving	Swerving	Swerving	Swerving	Swerving	Swerving	Swerving	Swerving
1.25	Jerky Driving	Jerky Driving	Jerky Driving	Swerving	Swerving	Swerving	Swerving	Swerving	Swerving	Swerving
1	Jerky Driving	Jerky Driving	Swerving	Swerving	Swerving	Swerving	Swerving	Swerving	Weaving	Weaving
0.75	Jerky Driving	Jerky Driving	Swerving	Swerving	Swerving	Weaving	Weaving	Weaving	Weaving	Weaving
0.5	Jerky Driving	Swerving	Swerving	Weaving	Weaving	Weaving	Weaving	Weaving	Normal Driving	Normal Driving
0.25	Jerky Driving	Weaving	Weaving	Normal Driving	Normal Driving	Normal Driving	Normal Driving	Normal Driving	Normal Driving	Normal Driving

the swerving output, which are mainly detected when the track's period is 5 s or above, although swerving can also occur at periods of less than 5 s if the amplitude is low. The 'brown blocks', which represent weaving, occur from the period of 2–10 s, provided the amplitude is in the range of 0.25–1 m. The 'grey blocks' (representing normal driving) occur across almost all time periods if the amplitude is very low, at 0.25 m. From the information revealed in the table, it can be concluded that as the input track's period increases and the amplitude decreases, the driving styles will be more likely to be detected as normal driving, which corresponds to the extreme situation that when the period becomes larger and the amplitude becomes smaller, the driving track nearly becomes a straight line. In contrast, when the period becomes shorter and amplitude larger, the styles will be more likely to be detected as jerky driving.

It is also clear that as the period increases (e.g. greater than 5 s), even if the amplitude is as high as 1.75 m, the driving track will be detected as swerving. This means that if the sinusoidal track's period is greater than 5 s, no jerky driving will be detected up to the maximum tested amplitude of 1.75 m. Meanwhile, as the period increases, the maximum amplitude allowing for the detection of normal driving increases.

Two issues need to be considered here. One is whether an increase in vehicle velocity will change the detected irregular driving type. In these experiments, the analysis has been based on a velocity of 85 km/h, but as the velocity increases, the O-indicator, related to the yaw angle values, will be affected by the velocity change. Thus, for an input vehicle track with the same period and amplitude, different velocities will result in different driving classification outputs. For example, when the velocity is 85 km/h for a track with an amplitude of 1.75 m and a period of 5 s, the driving style is detected as swerving. If the velocity is increased to 120 km/h, however, for the same track with amplitude of 1.75 m and period of 5 s, the driving style is detected as jerky driving. As is shown in Table 5.16, increases in the driving velocity will shift the predominance of the 'colour blocks' to the right as the velocity increases with the same period and amplitude values.

Another issue is whether the vehicle's width will affect the irregular driving detection results. In the simulation, the vehicle width is 1 m, which means that the maximum lateral displacement within the lane is 1.25 m. An increase in the vehicle width will only affect the maximum lateral displacement within the lane and is thus not relevant to the detection of irregular driving types. This is because from the analysis results, the detected driving styles are only related to the vehicle track's amplitude, period and velocity.

The episodes of irregular driving types were now tested to find out if the irregular driving detection algorithm can continuously detect different types of irregular driving. A 60 s episode was created to test the irregular driving detection algorithm. Figure 5.18 shows the trajectory of the designed episode. In order to make the data easy, the simulated data included manoeuvres at the speed of 85 km/h with weaving (sinusoidal wave with period $T = 5$ and amplitude $A = 0.5$ for 10 s), swerving ($T = 10$ and $A = 1.25$ for 10 s), jerky driving ($T = 2$ and $A = 0.75$ for 10 s) and normal driving ($T = 10$, $A = 0.25$ for 30 s).

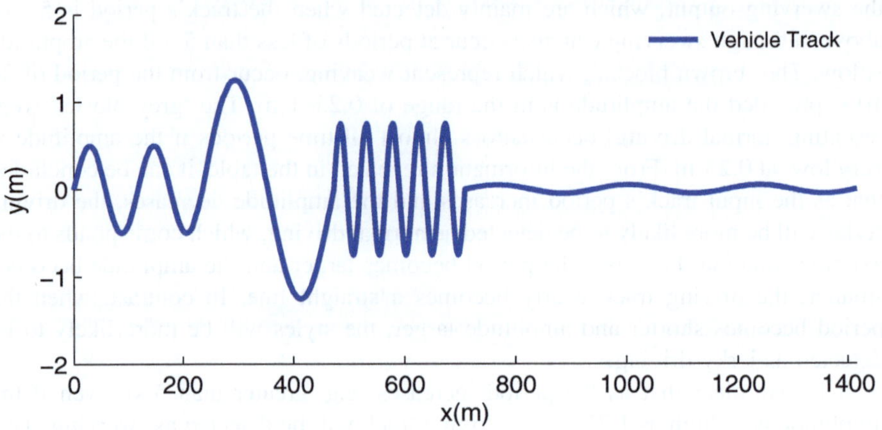

Fig. 5.18 Input episode with different driving styles

Table 5.17 Detection results for the episode

Output rate (Hz)	Time to first detection			
	Weaving (s)	Swerving (s)	Jerky driving (s)	Normal driving (s)
10	5	14.5	21.8	32.7
5	5	14.5	22	33
1	5	15	22	33

The time to first detection of driving style detection results for the test episode were output in 10, 5 and 1 Hz, see Table 5.17.

Since the algorithm uses the previous 5 s of data to judge the driving style, the first detection is at the 5 s output time. From the comparison of the irregular driving detection output at different rates, it is shown that if the output rate is higher, the detection of the specific types of driving style can be earlier. For example, if the output rate is 5 Hz, the first time in which jerky driving is detected in the 22 s time epoch, but if the rate is 10 Hz, jerky driving can be detected at 21.8 s, i.e. 0.2 s earlier than when using an output of 5 Hz.

Availability and correct detection rate are further parameters to evaluate the performance of the irregular driving detection algorithm. In general, although the episode test results show that every designed irregular driving style can be distinguished and detected, some problems still exist in the detection of irregular driving. Figure 5.19 is an example of unavailability in detection of irregular styles at 5 Hz. The driving classification indicator in the highlighted part forms the sorting rule CD = AD > AB = BC, which cannot output any driving styles. This kind of unavailability always happens at the point where the driving style is changing from one style to another. The availability and correct detection rate for the episode is

	Time	AB	BC	CD	AD	Detected Styles
47	28	0	30	50	20	jerky driving
48	28. 5	0	30	50	20	jerky driving
49	29	0	30	50	20	jerky driving
50	29. 5	0	27	50	23	jerky driving
51	30	0	24	50	26	jerky driving
52	30. 5	5	21	45	29	jerky driving
53	31	10	18	40	32	jerky driving
54	31. 5	15	15	35	35	N/A
55	32	20	12	30	38	jerky driving
56	32. 5	25	9	25	41	jerky driving
57	33	30	6	20	44	normal drivng
58	33. 5	35	3	15	47	normal drivng
59	34	40	0	10	50	normal drivng
60	34. 5	45	0	5	50	normal drivng
61	35	50	0	0	50	normal drivng
62	35. 5	50	0	0	50	normal drivng

Fig. 5.19 An example of unavailability in detection of irregular driving

Table 5.18 The availability and correct detection rate for the episode analysis with respect to output rate

Output rate (s)	Availability (%)	Correct detection rate (%)
0.1	94.19	97.82
0.5	98.19	98.20
1	98.21	100

shown in Table 5.18 and it is clear that if the output rate of the system is higher, the correct detection rate and availability rate will decrease. Thus, choosing the proper output rate for the required situation is critical.

5.3 Simulated Measurement and Results Analysis

Besides the reference trajectory and dynamic parameters generated for the scenarios, the simulated GPS and INS measurement data is generated following the reference trajectory. The simulated measurements feed the filters and motion models with integrated algorithms to compute estimated *omega* and *d* data for O-indicator and D-indicator generation and then interpretation by Fuzzy Inference System (FIS) to calculate risk type indicators and driving classification indicators for irregular driving detection scenarios.

High accuracy positioning related results are a basic requirement for the systems to identify irregular driving. In this subsection, the simulated GPS measurements and INS measurements for the vehicle positioning information and dynamic motion

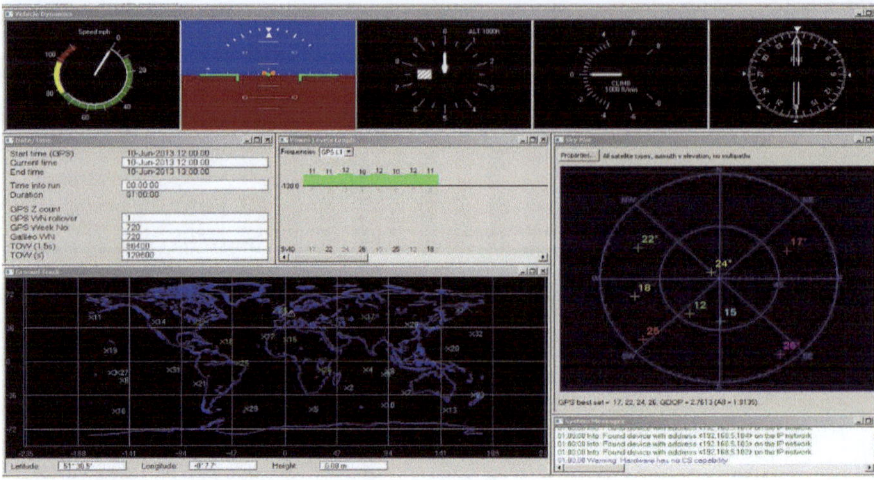

Fig. 5.20 GNSS simulator display

information are generated along the predefined scenario routes to analyse the performance of the designed sensor integration models and determine the required positioning accuracy for the irregular driving detection algorithm.

The GPS measurement is created by a Leica GR10 receiver with a Spirent GSS8000 GNSS simulator. It is used to simulate the constellation on 10-June-2013 with GPS only in London. The output rate of simulated GPS measurements is 10 Hz. Figure 5.20 shows the GNSS simulator display, with the simulated vehicle's position, dynamic parameters and the sky plot of satellite numbers and distributions at the simulated time epoch. The 10 Hz INS measurement including the acceleration and yaw rate is simulated by the Matlab.

The results for the straight lane and curved lane scenarios are discussed in the following sub-section. The accuracy for the EKFCV, EKFCA, PFCV and PFCA models are compared by scenarios, and the driving classification indicators calculated based on each fusion model are analysed.

5.3.1 *Straight Scenarios*

By comparing the integrated positioning results from EKFCV, EKFCA, PFCV and PFCA model estimations, in general, the EKF filters produce more noise, while the PF filter estimations are comparatively smooth (see Figs. 5.21, 5.22, 5.23 and 5.24). The 2dRMS vehicle positioning accuracy based on EKFCV, EKFCA, PFCV and PFCA models for S1, S2, S3, and S4 are shown in Table 5.19, which indicates that all of the precise positioning estimation models have achieved an accuracy better than 1 m in all scenarios. Among these designed positioning estimation models,

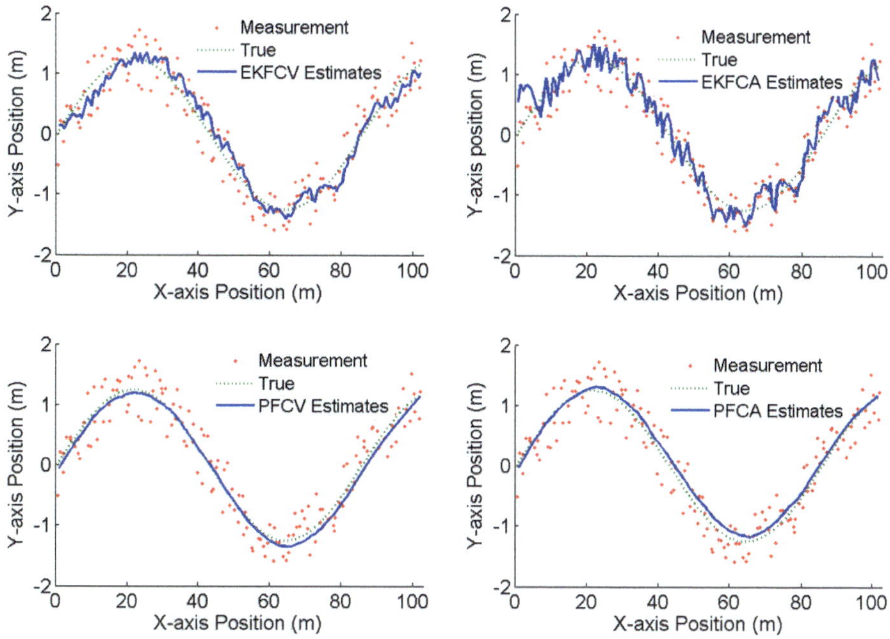

Fig. 5.21 Integrated estimation positioning results for S1

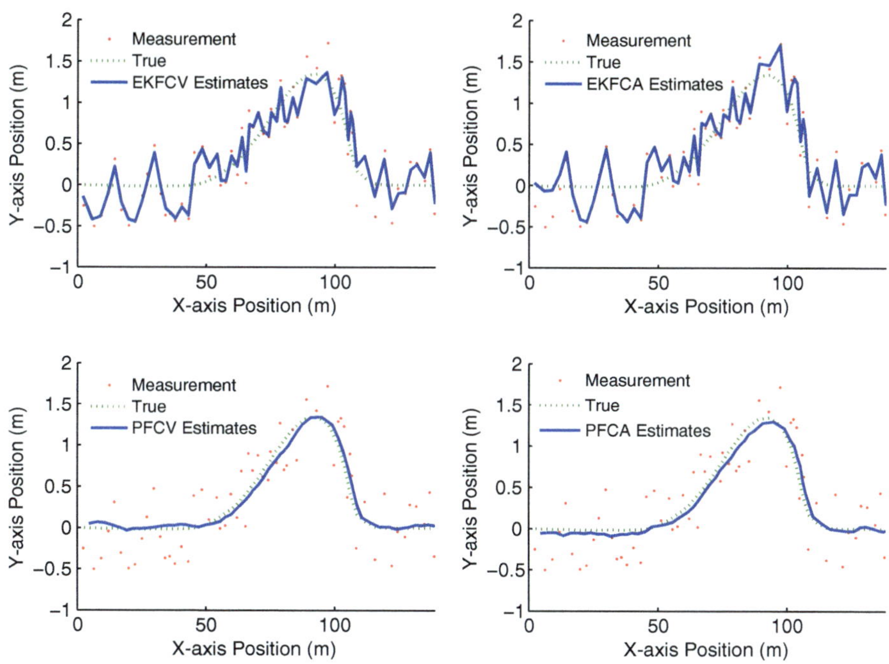

Fig. 5.22 Integrated estimation positioning results for S2

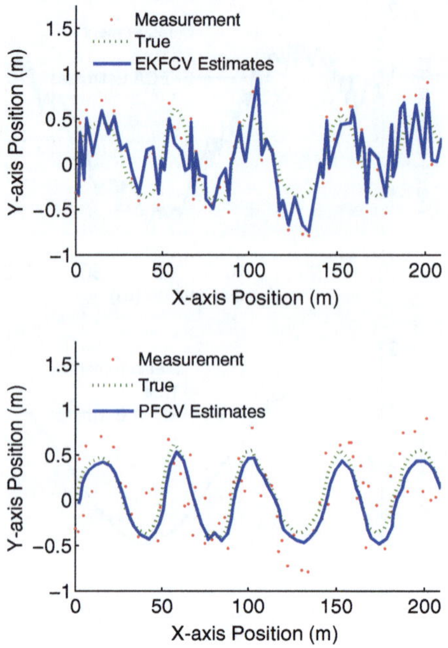

Fig. 5.23 Integrated estimation positioning results for S3

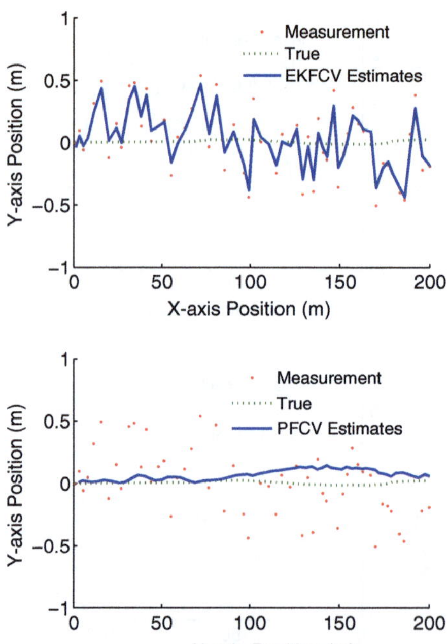

Fig. 5.24 Integrated estimation positioning results for S4

Table 5.19 Accuracy comparison of model estimated results for straight scenarios

Straight scenarios accuracy 2dRMS comparison (m)	EKFCV	EKFCA	PFCV	PFCA
S1 weaving	0.3925	0.3956	0.2621	0.2142
S2 swerving	0.5624	0.5820	0.3453	0.3410
S3 jerky driving	0.5426	0.5002	0.3801	0.3713
S4 normal driving	0.5125	0.4826	0.3920	0.3815

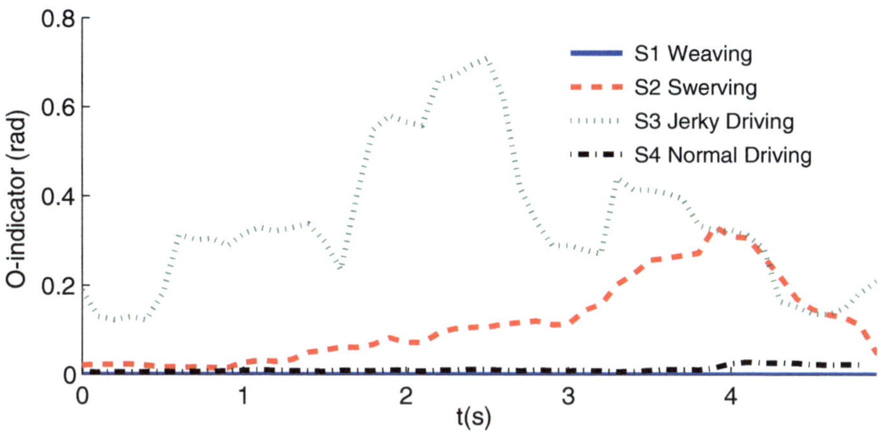

Fig. 5.25 EKFCV estimated O-indicator comparison

PFCV and PFCA models achieve an accuracy better than 0.5 m in all of the scenarios, while EKFCV and EKFCA achieve an accuracy of between 0.5 and 1 m in some scenarios.

The analysis of the irregular driving detection algorithm based on model estimated results for the straight scenarios is discussed below.

EKFCV

The O-indicator and D-indicator from EKFCV estimation results are shown in Figs. 5.25 and 5.26. The corresponding risk type indicator is shown in Fig. 5.27. Tables 5.20 and 5.21 show the statistical results of the risk types indicator and driving classification indicator based on the EKFCV model. From the calculation and comparison results of the AB, BC, CD and AD values in Table 5.21, the S1 weaving feature is easily detected, since its driving classification indicator forms the rule AB > BC > AD > CD, specified for the weaving scenario. The correct detection also goes for S3 jerky driving and S4 normal driving. S2 swerving, however, is detected as jerky driving.

EKFCA

O-indicator and D-indicator results from the EKFCA fusion model are shown in Figs. 5.28 and 5.29. The corresponding risk type indicator is shown in Fig. 5.30.

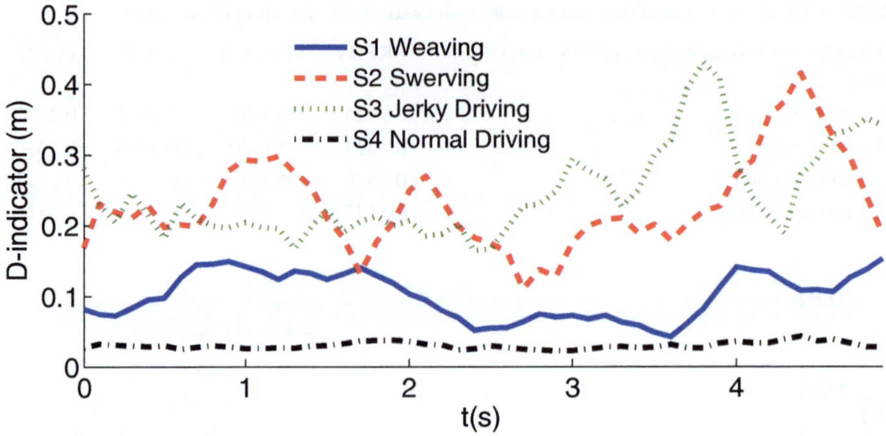

Fig. 5.26 EKFCV estimated D-indicator comparison

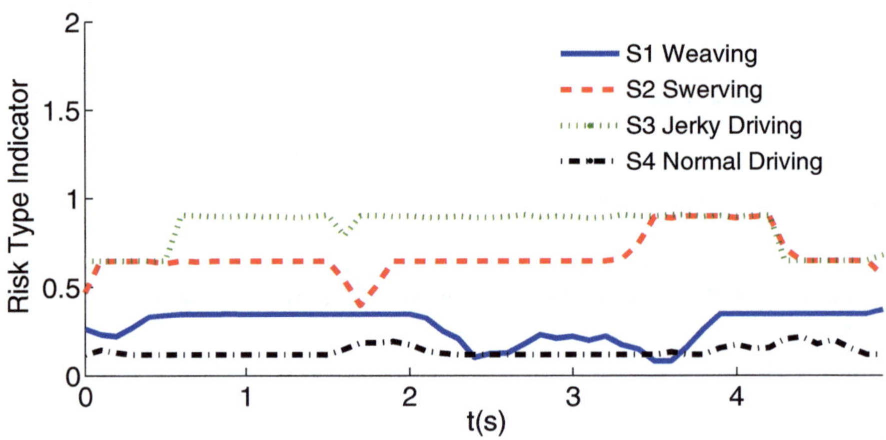

Fig. 5.27 EKFCV estimated risk type indicator comparison

Table 5.20 Statistics of number of points in risk types for straight lane scenarios based on EKFCV

Number of points	Risk types for EKFCV results			
Scenarios	A	B	C	D
S1 weaving	17	33	0	0
S2 swerving	0	2	40	8
S3 jerky driving	0	0	14	36
S4 normal driving	50	0	0	0

Table 5.21 Driving classification indicators for straight scenarios based on EKFCV

Scenarios	AB	BC	CD	AD	Judgement
S1 weaving	50	33	0	17	Weaving
S2 swerving	2	42	48	8	Jerky driving
S3 jerky driving	0	14	50	36	Jerky driving
S4 normal driving	50	0	0	50	Normal driving

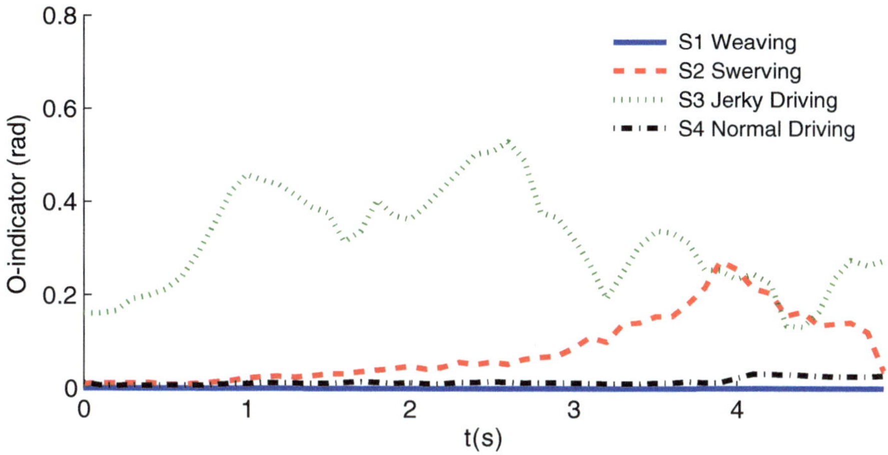

Fig. 5.28 EKFCA estimated O-indicator comparison

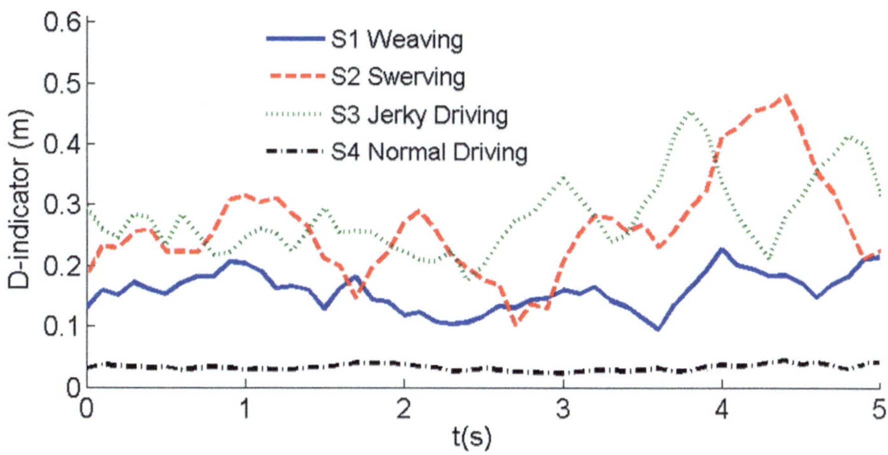

Fig. 5.29 EKFCA estimated D-indicator comparison

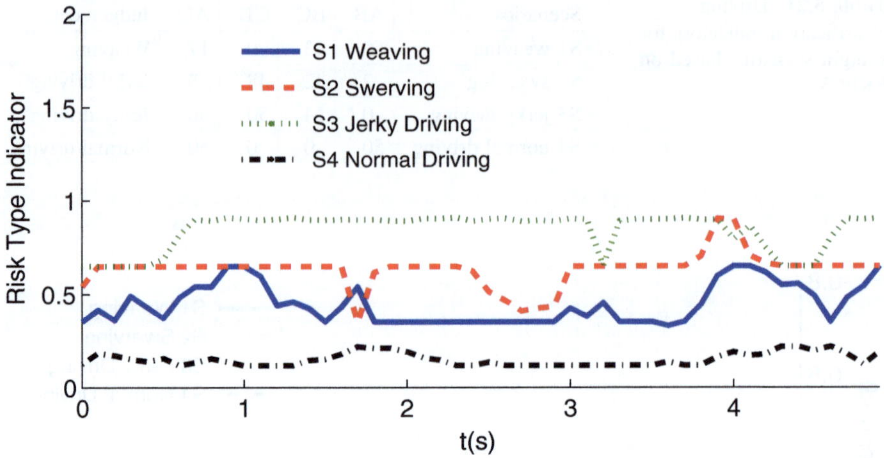

Fig. 5.30 EKFCA estimated risk type indicator comparison

Table 5.22 Statistics of number of points in risk types for straight lane scenarios based on EKFCA

Number of points	Risk types for EKFCA results			
Scenarios	A	B	C	D
S1 weaving	0	36	14	0
S2 swerving	0	5	43	2
S3 jerky driving	0	0	12	38
S4 normal driving	50	0	0	0

Table 5.23 Driving classification indicators for straight scenarios based on EKFCA

Scenarios	AB	BC	CD	AD	Judgement
S1 weaving	36	50	14	0	Swerving
S2 swerving	5	48	45	2	Swerving
S3 jerky driving	0	12	50	38	Jerky driving
S4 normal driving	50	0	0	50	Normal driving

Table 5.22 shows the statistical results of the risk types indicator based on the EKFCA. Table 5.23 presents the driving classification indicators for straight lane scenarios based on EKFCA. From a comparison of the results in Table 5.23, it can be seen that all the features in all the scenarios can be identified except weaving, which is wrongly-detected as swerving.

PFCV

O-indicator and D-indicator from the PFCV model estimated results are shown in Figs. 5.31 and 5.32. The corresponding risk type indicator values are shown in Fig. 5.33. Table 5.24 presents the statistical results of risk types based on the PFCV. Table 5.25 shows the driving classification indicators for the straight lane scenarios based on PFCV. In Table 5.25, the sorting of parameters in S1, which follows

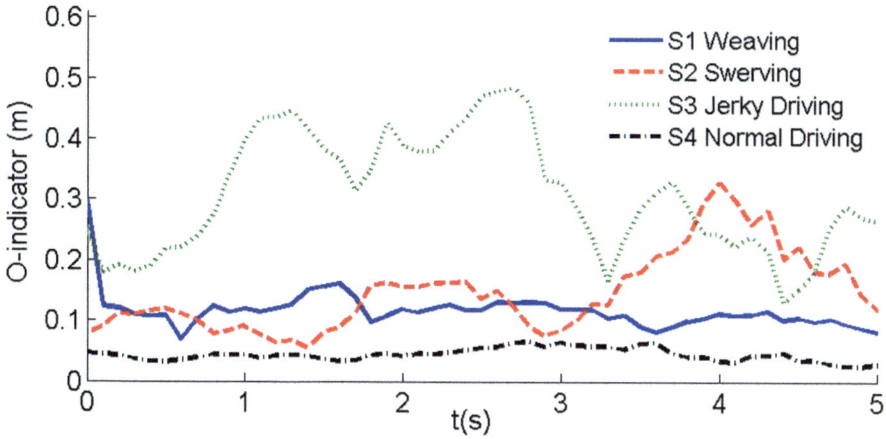

Fig. 5.31 PFCV estimated O-indicator comparison

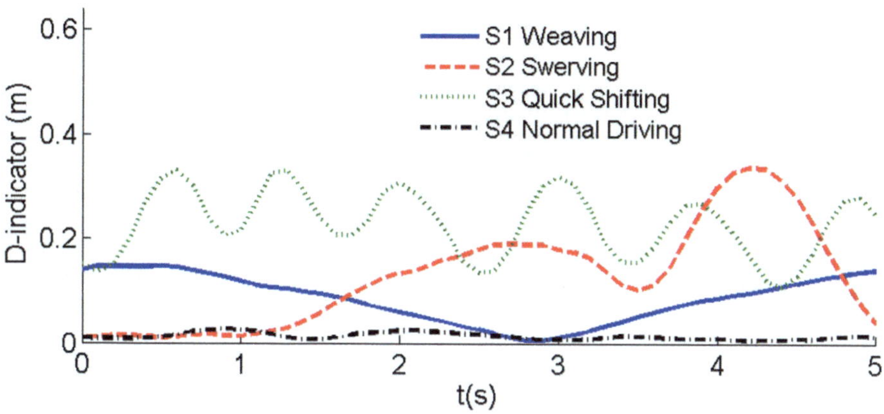

Fig. 5.32 PFCV estimated D-indicator comparison

AB > BC > AD > CD, is identified as weaving; S2, BC > CD > AB > AD, is identified as swerving; S3, CD > AD > BC > AB, is identified as jerky driving; S4, AB > AD > BC > CD, is identified as normal driving. The results show that the PFVC model results lead to the correct identification of all the straight lane scenarios.

PFCA

The O-indicator and D-indicator results are shown in Figs. 5.34 and 5.35. The PFCA estimated risk type indicator values are shown in Fig. 5.36. Tables 5.26 and 5.27 show the statistical results of risk types and driving classification indicators based on the PFCA. Similar to PFCV, the PFCA based identification is successful in all of the straight lane scenarios. From Table 5.27, it is shown that AB > BC > AD > CD is identified as weaving in S1; BC > CD > AB > AD is

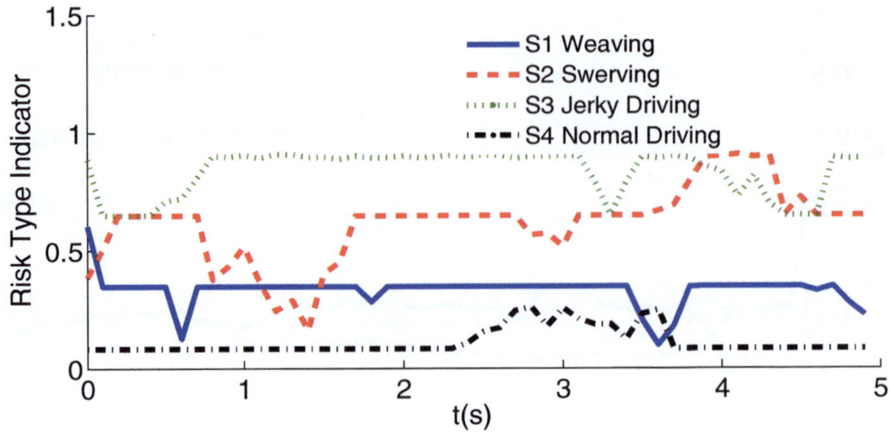

Fig. 5.33 PFCV estimated risk type indicator comparison

Table 5.24 Statistics of number of points in risk types for straight lane scenarios based on PFCV

Number of points	Risk types for PFCV results			
Scenarios	A	B	C	D
S1 weaving	5	44	1	0
S2 swerving	2	7	36	5
S3 jerky driving	0	0	14	36
S4 normal driving	48	2	0	0

Table 5.25 Driving classification indicators for straight scenarios based on PFCV

Scenarios	AB	BC	CD	AD	Judgement
S1 weaving	49	45	1	5	Weaving
S2 swerving	9	43	41	7	Swerving
S3 jerky driving	0	14	50	36	Jerky driving
S4 normal driving	50	2	0	48	Normal driving

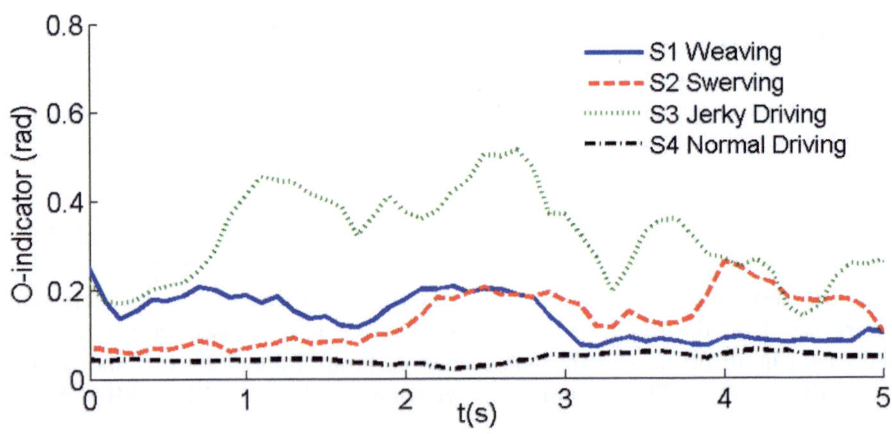

Fig. 5.34 PFCA estimated O-indicator comparison

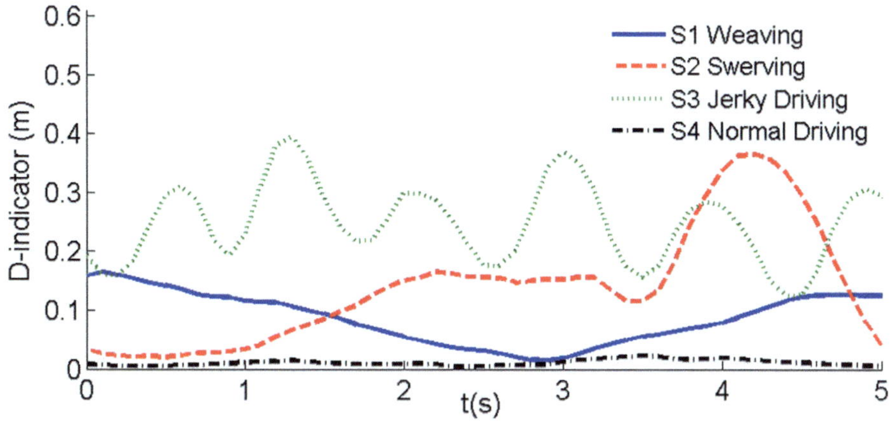

Fig. 5.35 PFCA estimated D-indicator comparison

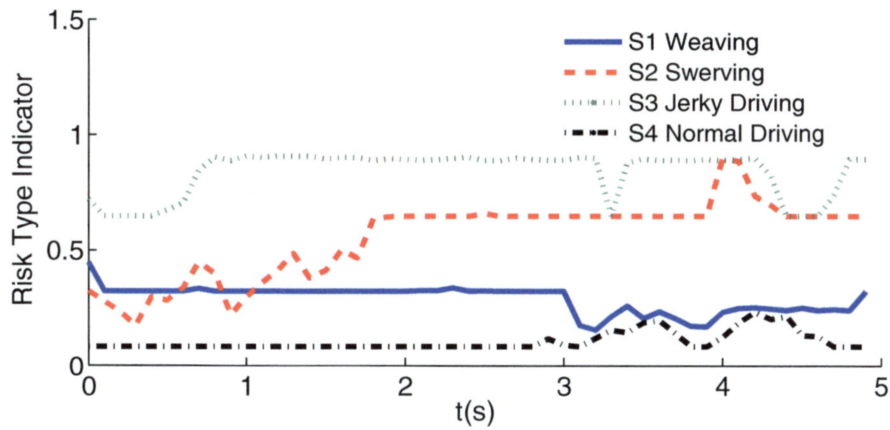

Fig. 5.36 PFCA estimated risk type indicator comparison

Table 5.26 Statistics of number of points in risk types for straight lane scenarios based on PFCA

Number of points	Risk types for PFCA results			
Scenarios	A	B	C	D
S1 weaving	14	36	0	0
S2 swerving	3	14	31	2
S3 jerky driving	0	0	12	38
S4 normal driving	50	0	0	0

Table 5.27 Driving classification indicators for straight scenarios based on PFCA

Scenarios	AB	BC	CD	AD	Judgement
S1 Weaving	50	36	0	14	Weaving
S2 Swerving	17	45	33	5	Swerving
S3 jerky driving	0	12	50	38	Jerky driving
S4 normal driving	50	0	0	50	Normal driving

identified as swerving in S2; CD > AD > BC > AB is identified as jerky driving in S3; AB > AD > BC > CD is identified as normal driving in S4.

5.3.2 Curved Scenarios

Figures 5.37, 5.38, 5.39, 5.40 and 5.41 present graphically the simulation results for the curved lane scenarios. The 2dRMS accuracy based on the EKFCTRV, EKFCTRA, PFCTRV and PFCTRA model estimated results for S5, S6a, S6b, S7 and S8 are shown in Table 5.28, which also indicates that all of the precise positioning estimation models have achieved an accuracy better than 1 m. Among these designed positioning estimation models, only the PFCTRA model achieves an accuracy better than 0.5 m in all of the scenarios, while the other models such as EKFCTRV, EKFCTRA and PFCTRV achieve an accuracy of between 0.5 and 1 m in some scenarios in the simulation.

The filter estimated results for irregular driving detection on the curved scenarios are discussed below.

EKFCTRV

The O-indicator and D-indicator from the EKFCTRV estimation are shown in Figs. 5.42 and 5.43. The corresponding risk type indicator values are shown in Fig. 5.44. Tables 5.29 and 5.30 show the statistical results for the risk types and driving classification indicators based on the EKFCTRV estimations. From the

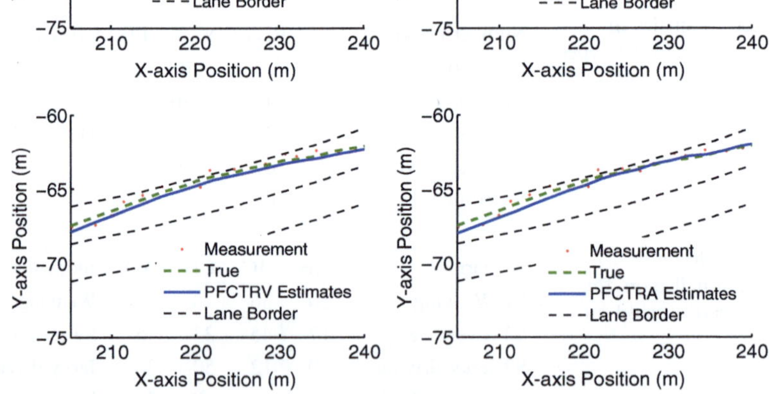

Fig. 5.37 Integrated estimation positioning results for S5

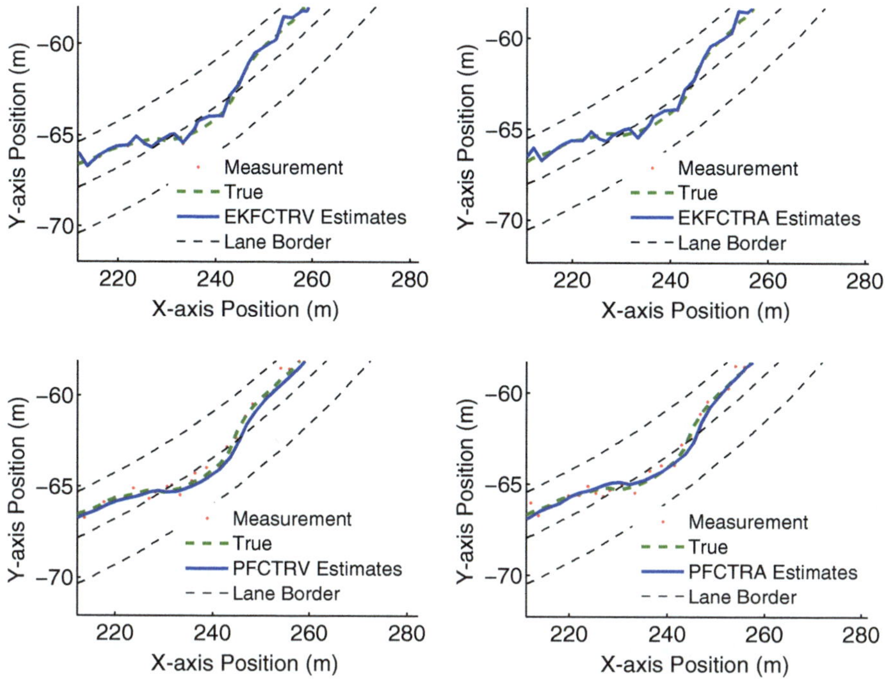

Fig. 5.38 Integrated estimation positioning results for S6a swerving (over-turning)

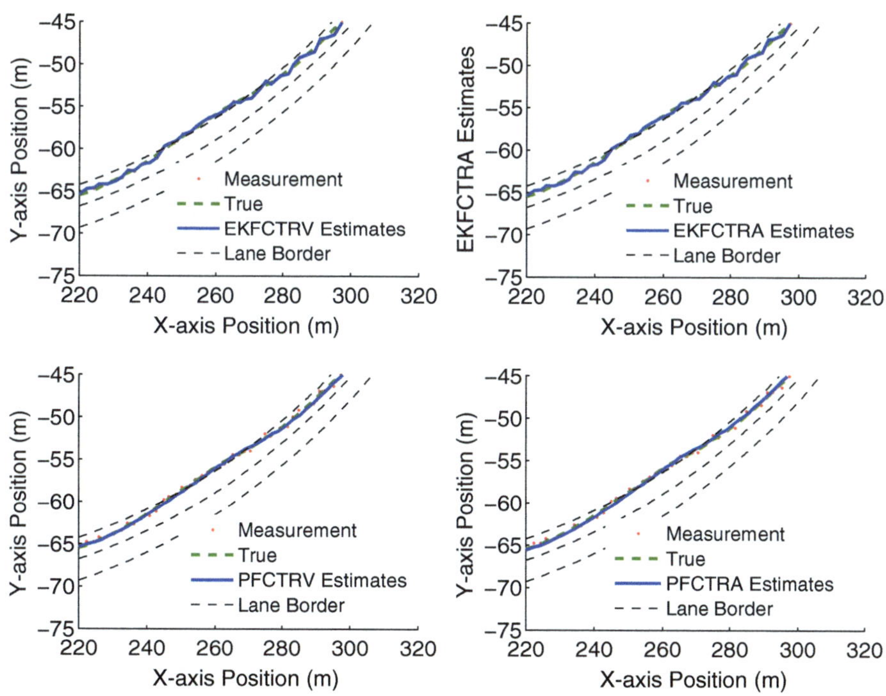

Fig. 5.39 Integrated estimation positioning results for S6b swerving (under-turning)

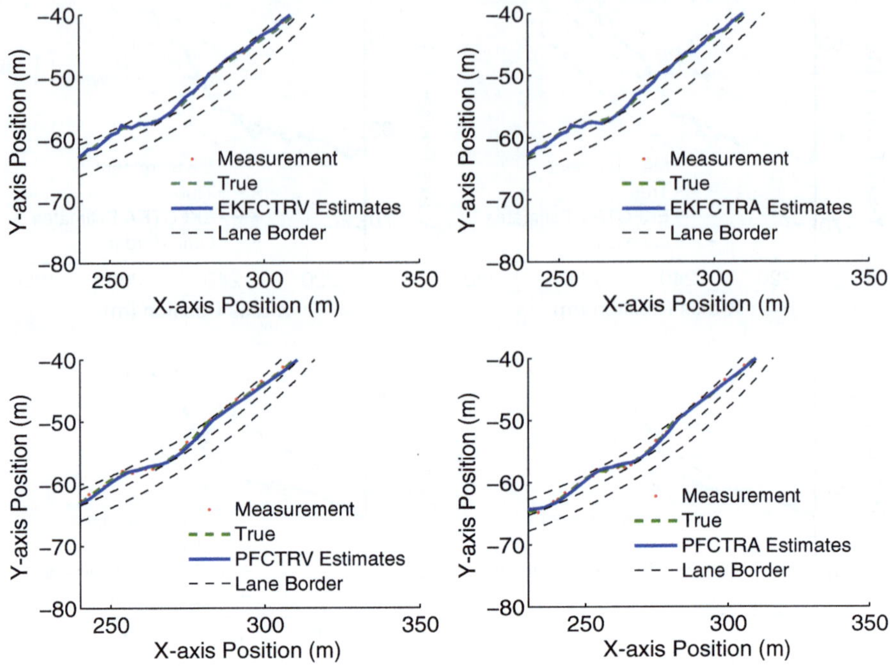

Fig. 5.40 Integrated estimation positioning results for S7

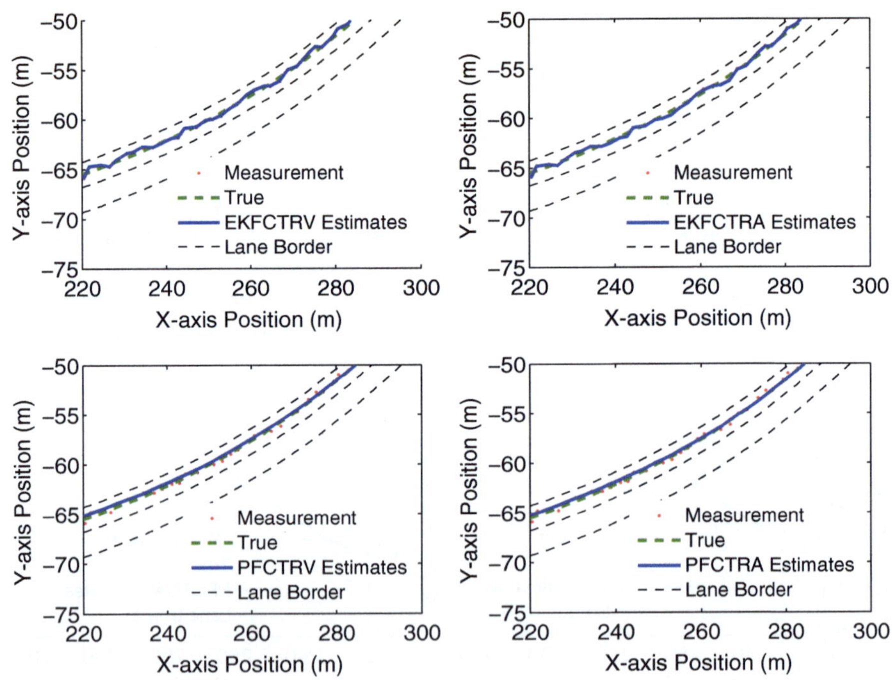

Fig. 5.41 Integrated estimation positioning results for S8

Table 5.28 Accuracy comparison of model estimated results for curved lane scenarios

Curved scenarios accuracy 2dRMS comparison (m)	EKFCTRV	EKFCTRA	PFCTRV	PFCTRA
S5 weaving on curve	0.561	0.451	0.410	0.391
S6a swerving (over-turning) on curve	0.545	0.478	0.390	0.374
S6b swerving (under-turning) on curve	0.427	0.411	0.371	0.361
S7 jerky driving on curve	0.742	0.686	0.544	0.432
S8 normal driving on curve	0.483	0.441	0.396	0.374

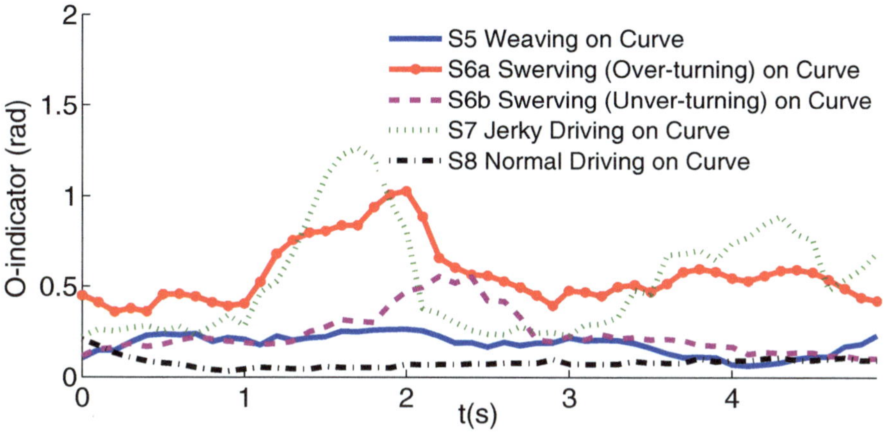

Fig. 5.42 EKFCTRV estimated O-indicator comparison

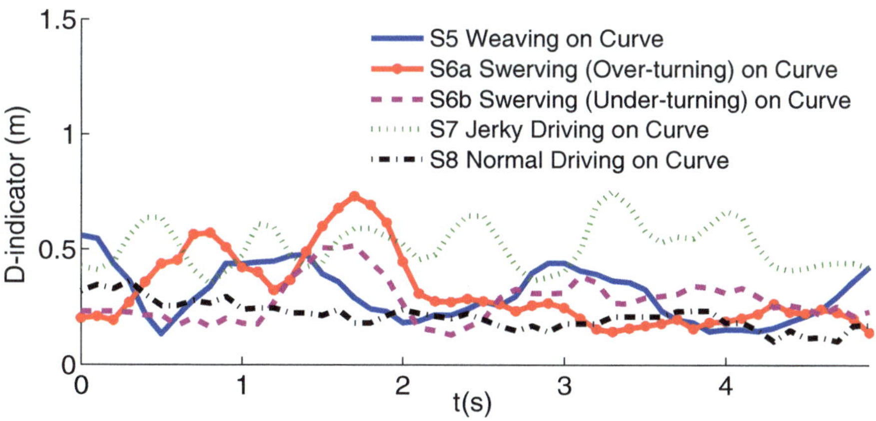

Fig. 5.43 EKFCTRV estimated D-indicator comparison

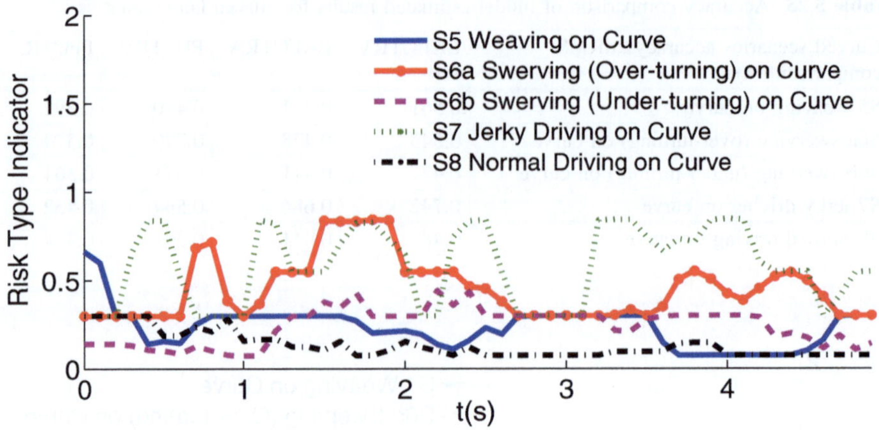

Fig. 5.44 EKFCTRV estimated risk type indicator comparison

Table 5.29 Statistics of number of points in risk types for curved based on EKFCTRV

Number of points	Risk types for EKFCTRV results			
Scenarios	A	B	C	D
S5 weaving on curve	14	34	1	1
S6a swerving (over-turning) on curve	0	27	15	7
S6b swerving (under-turning) on curve	15	35	0	0
S7 jerky driving on curve	0	16	12	22
S8 normal driving on curve	35	15	0	0

Table 5.30 Driving classification indicators for curved scenarios based on EKFCTRV

Scenarios	AB	BC	CD	AD	Judgement
S5 weaving on curve	48	35	2	15	Weaving on curve
S6a swerving (over-turning) on curve	27	42	22	7	Swerving on curve
S6b swerving (under-turning) on curve	50	35	0	15	Weaving on curve
S7 jerky driving on curve	16	28	34	22	Jerky driving on curve
S8 normal driving on curve	50	15	0	35	Normal driving on curve

results, it is evident that the EKFCTRV results have wrongly detected S6b Swerving (under-turning) as weaving on a curve. The other scenarios, however, are detected correctly.

EKFCTRA

The O-indicator and D-indicator results from the EKFCTRA model estimation are shown in Figs. 5.45 and 5.46. The corresponding risk type indicator values are

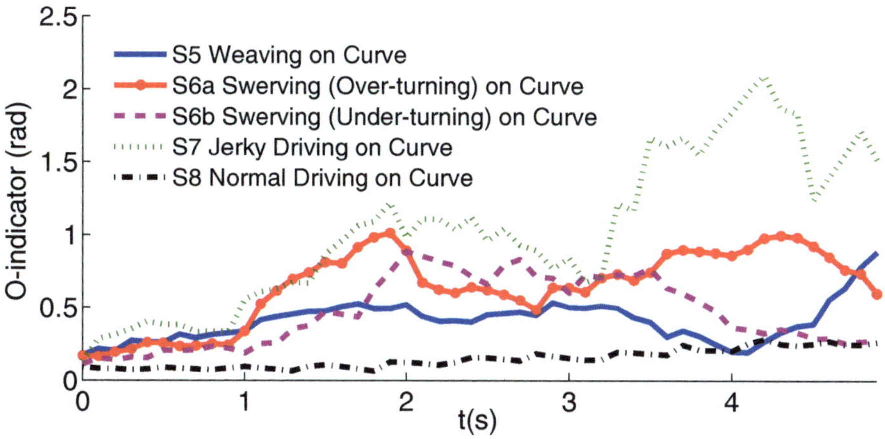

Fig. 5.45 EKFCTRA estimated O-indicator comparison

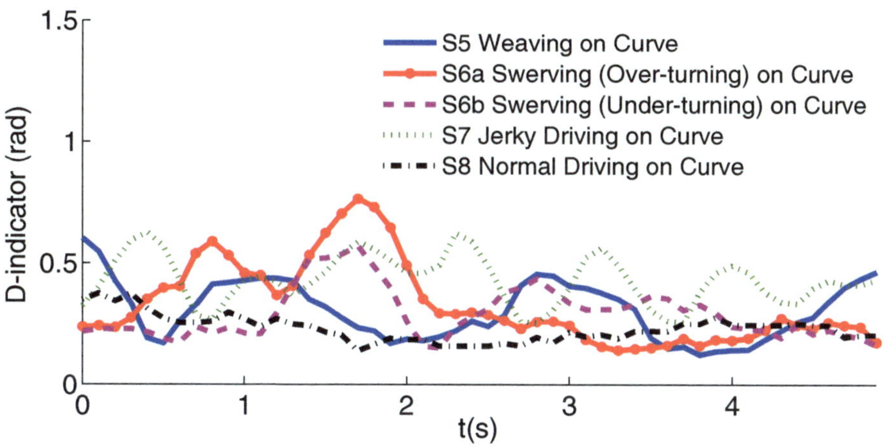

Fig. 5.46 EKFCTRA estimated D-indicator comparison

shown in Fig. 5.47. Tables 5.31 and 5.32 show the calculated statistical results of risk types and driving classification indicator based on the EKFCTRA estimated parameters. From Table 5.32, it is evident that the EKFCTRA results have some missed detections in some scenarios. The S5 weaving on a curve feature is wrongly detected as swerving on a curve. Also for S8, normal driving on a curve is detected as weaving on a curve. Swerving and jerky driving scenarios are identified correctly.

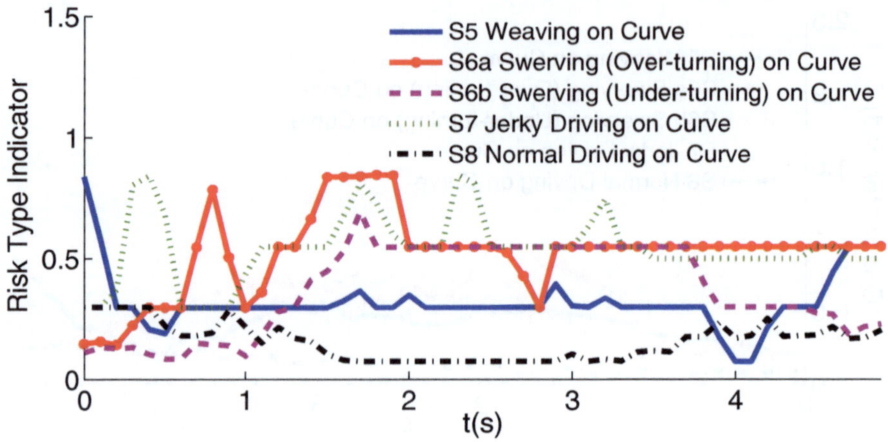

Fig. 5.47 EKFCTRA estimated risk type indicator comparison

Table 5.31 Statistics of number of points in risk types for curved lane scenarios based on EKFCTRA

Number of points	Risk types for EKFCTRA results			
Scenarios	A	B	C	D
S5 weaving on curve	2	43	4	1
S6a swerving (over-turning) on curve	2	9	32	7
S6b swerving (under-turning) on curve	10	18	21	1
S7 jerky driving on curve	0	8	33	9
S8 normal driving on curve	23	27	0	0

Table 5.32 Driving classification indicators for straight scenarios based on EKFCTRA

Scenarios	AB	BC	CD	AD	Judgement
S5 weaving on curve	45	47	5	3	Swerving on curve
S6a swerving (over-turning) on curve	11	41	39	9	Swerving on curve
S6b swerving (under-turning) on curve	28	39	22	11	Swerving on curve
S7 jerky driving on curve	8	41	42	9	Jerky driving on curve
S8 normal driving on curve	50	27	0	23	Weaving on curve

PFCTRV

The O-indicator and D-indicator results are shown in Figs. 5.48 and 5.49. The PFCTRV estimated risk type indicator results are shown in Fig. 5.50. Tables 5.33 and 5.34 indicate the statistical results of risk types and driving classification indicators based on the PFCTRV. From the results, it is shown that in S6a Swerving (over-turning) is wrongly detected as weaving, but all other scenarios are identified correctly.

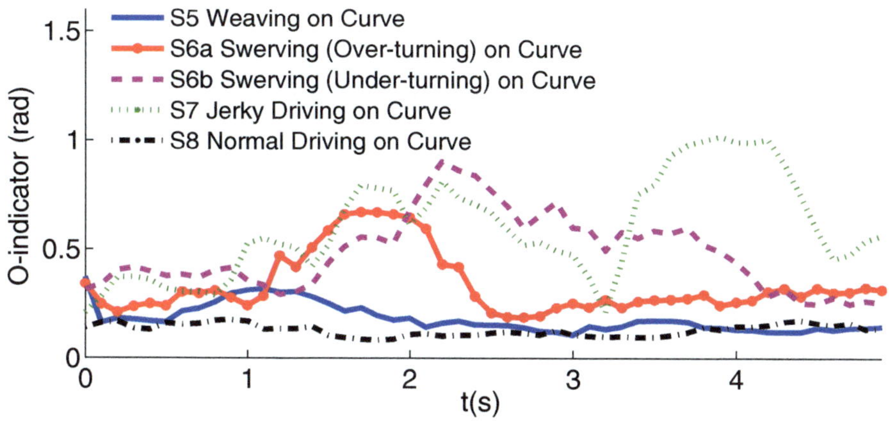

Fig. 5.48 PFCTRV estimated O-indicator comparison

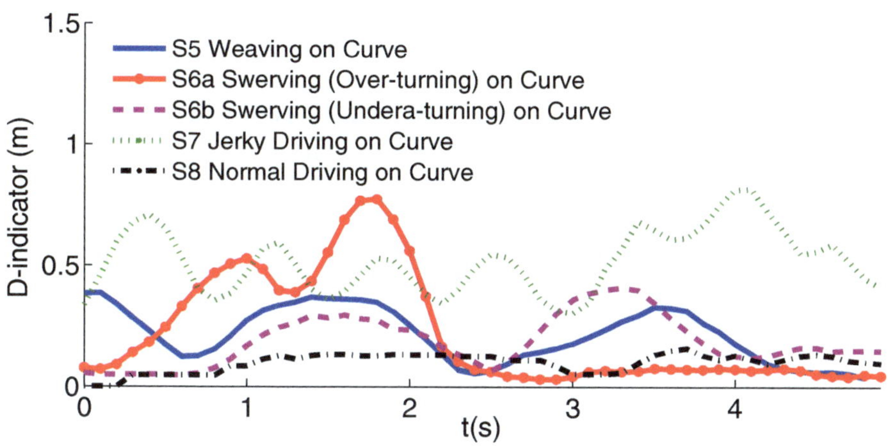

Fig. 5.49 PFCTRV estimated D-indicator comparison

PFCTRA

The O-indicator and D-indicator from the PFCTRA estimated results are shown in Figs. 5.51 and 5.52. The PFCTRA estimated risk type indicator values are shown in Fig. 5.53. Tables 5.35 and 5.36 present the statistical results of risk types and driving classification indicators based on PFCTRA. From the results, it is evident that the PFCTRA based results have correctly detected all scenarios and it also achieves the best estimation of the values in the risk type indicator, and the driving style classification indicators are closest to the 'true data' FIS output.

To sum up, from the filter and motion model estimated vehicle positioning, in the straight scenarios, PFCV and PFCA exhibit an accuracy better than 0.5 m in all scenarios while, in the curved scenarios, only PFCTRA exhibits an accuracy better

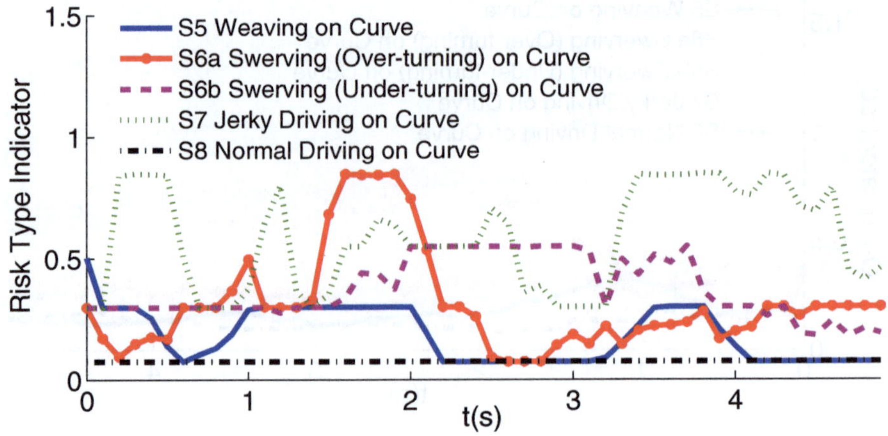

Fig. 5.50 PFCTRV estimated risk type indicator comparison

Table 5.33 Statistics of number of points in risk types for curved lane scenarios based on PFCTRV

Number of points	Risk types for PFCTRV results			
Scenarios	A	B	C	D
S5 weaving on curve	24	26	0	0
S6a swerving (over-turning) on curve	9	33	2	6
S6b swerving (under-turning) on curve	0	34	16	0
S7 jerky driving on curve	0	16	13	21
S8 normal driving on curve	50	0	0	0

Table 5.34 Driving classification indicators for straight scenarios based on PFCTRV

Scenarios	AB	BC	CD	AD	Judgement
S5 weaving on curve	50	26	0	24	Weaving on curve
S6a swerving (over-turning) on curve	42	35	8	15	Weaving on curve
S6b swerving (under-turning) on curve	34	50	16	0	Swerving on curve
S7 jerky driving on curve	16	29	34	21	Jerky driving on curve
S8 normal driving on curve	50	0	0	50	Normal driving on curve

than 0.5 m in all scenarios. In addition, the PF model exhibits a higher accuracy than the EKF model and the CA model exhibits a higher accuracy than the CV model during the simulation. Thus, PFCA and PFCTRA achieve the highest positioning accuracy in their scenarios, respectively.

The performance of the model estimated results for irregular driving detection are discussed based on the positioning accuracy levels. The models with an

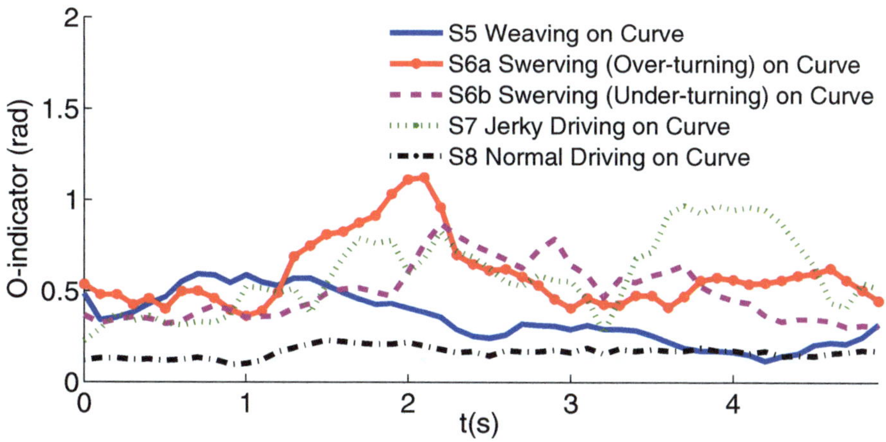

Fig. 5.51 PFCTRA estimated O-indicator comparison

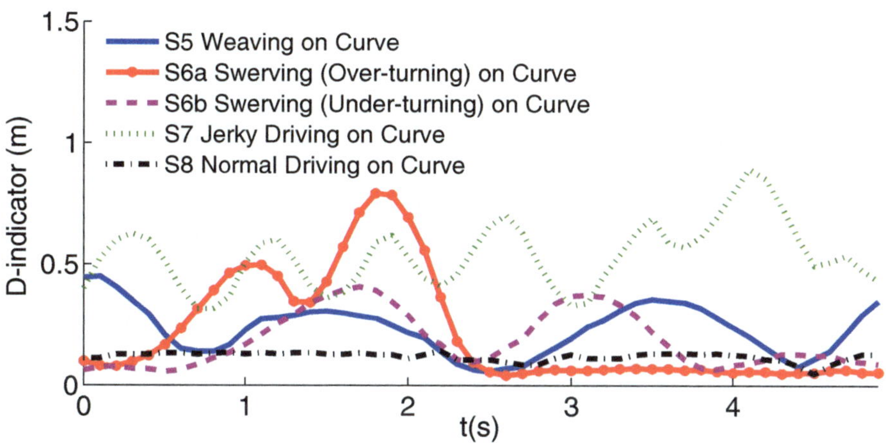

Fig. 5.52 PFCTRA estimated D-indicator comparison

estimation accuracy of above 0.5 m are EKFCV, EKFCTRV, EKFCA, EKFCRA and PFCTRV. EKFCV and EKFCTRV results, in addition, wrongly detected swerving/under-turning as jerky driving or weaving, while the EKFCA and EKFCTRA model results both wrongly detected weaving/weaving on a curve as swerving/swerving on a curve. Furthermore, the EKFCTRA estimated results also wrongly detected normal driving as weaving on a curve. The PFCTRV model results incorrectly detected swerving on a curve as weaving on a curve. It is obvious that none of the model estimation results above can be used for the correct detection of irregular driving scenarios due to their low positioning accuracy.

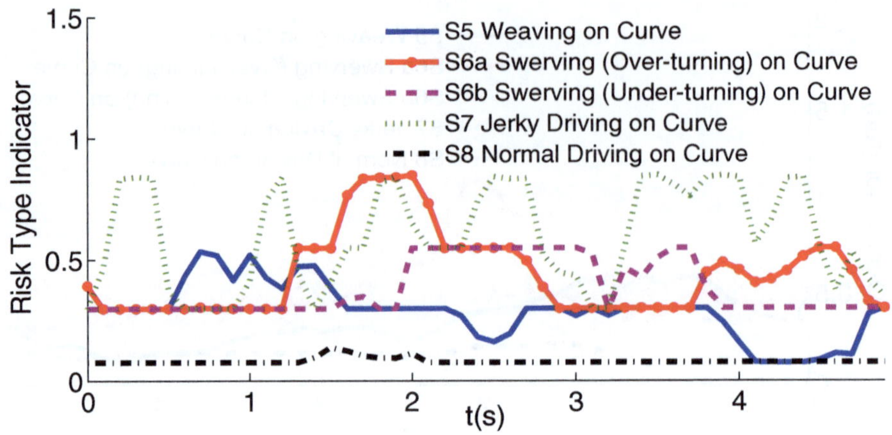

Fig. 5.53 PFCTRA estimated risk type indicator comparison

Table 5.35 Statistics of number of points in risk types for curved lane scenarios based on PFCTRA

Number of points	Risk types for PFCTRA results			
Scenarios	A	B	C	D
S5 weaving on curve	7	38	5	0
S6a swerving (over-turning) on curve	0	28	16	6
S6b swerving (under-turning) on curve	0	34	16	0
S7 jerky driving on curve	0	17	11	22
S8 normal driving on curve	50	0	0	0

Table 5.36 Driving classification indicators for straight lane scenarios based on PFCTRA

Scenarios	AB	BC	CD	AD	Judgement
S5 weaving on curve	45	43	5	7	Weaving on curve
S6a swerving (over-turning) on curve	28	44	22	6	Swerving on curve
S6b swerving (under-turning) on curve	34	50	16	0	Swerving on curve
S7 jerky driving on curve	17	28	33	22	Jerky driving on curve
S8 normal driving on curve	50	0	0	50	Normal driving on curve

PFCV, PFCA and PFCTRA model estimations, however, exhibit an accuracy better than 0.5 m, and all three of these models correctly detected the different irregular driving scenarios. From the simulation, it can therefore be concluded that a positioning accuracy of at least 0.5 m is required for the irregular driving detection algorithm.

In the next section, a field test is carried out to validate the designed lane level high accuracy positioning based irregular driving detection algorithm. Since the

PFCA and PFCTRA models achieve the best positioning estimations in the simulation it is these that will be applied to the true GPS data collected on the motorway.

5.4 Field Test and Results Analysis

The main objective of the real road test experiment is to evaluate the performance of the developed irregular driving detection algorithm in an actual situation. In this section, the irregular driving is carried out on a real road in a vehicle equipped with INS and GPS, with the high grade INS data being collected as the reference with which to compare the PFCA/PFCTRA model estimated RTK GPS measurements for the irregular driving detection performance analysis.

5.4.1 Data Collection Scheme Design

The designed driving routes for the experiment are on the M4 motorway from Ravenscourt Park to Heathrow Terminal 3 and then back to Imperial College London. Figure 5.54 shows the landscape of the driving routes.

The irregular driving styles are conducted on the two routes captured. For the first route in the east-west direction, data was captured for 7 min, involving weaving and jerky driving on a straight lane and jerky driving on a curved lane.

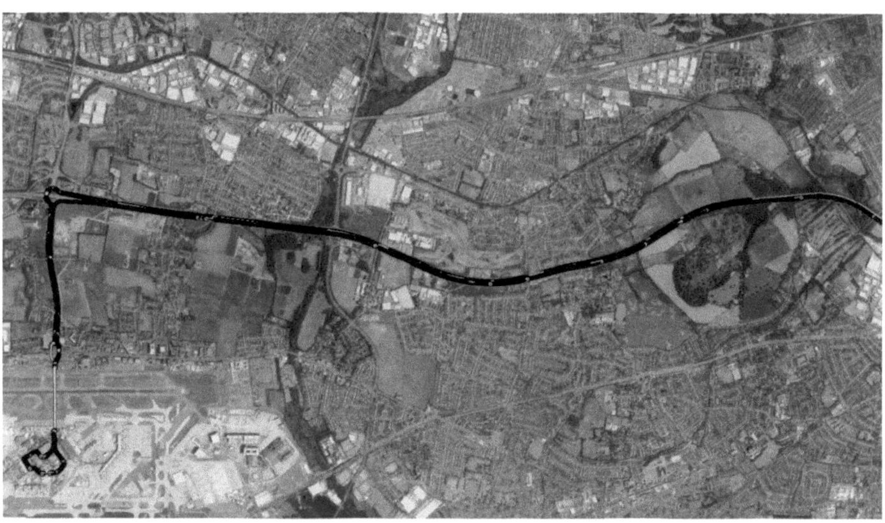

Fig. 5.54 Landscape of the driving route between ravenscourt park and heathrow terminal 3

Table 5.37 Definition of sessions

Session name	Start time (UTC)	End time (UTC)	Driving type
Session 1	15:51:14.0	15:51:20.8	Weaving on straight
Session 2	15:51:41.5	15:51:50.4	Jerky driving on curve
Session 3	15:57:13.9	15:57:23.5	Jerky driving on straight
Session 4	16:12:10.9	16:12:16.1	Swerving on curve
Session 5	16:12:36.4	16:12:43.4	Weaving on curve

For the second route in the west-east direction, data was captured for 3 min, involving swerving and weaving on a curved lane. In order to distinguish different types of irregular driving a total of five sessions are defined in Table 5.37. In order to ensure the safety of the experiment and high quality RTK measurements from the GPS, all the sessions were carried out in an open area with no other vehicles passing by and no trees on the side of the motorway. Since the experimental vehicle was quite old and potentially dangerous for some types of manoeuvres, swerving was attempted only once.

For the irregular driving data collection, the vehicle was driven at speeds ranging from 70 to 120 km/h and the data was collected at 10 Hz. The installation of the equipment is shown in Fig. 5.55. The specifics of the equipment used in the experiment were as follows.

Fig. 5.55 On board equipment

- Leica Viva GNSS GS15 receiver for real-time GPS data collection, including the position and speed of the vehicle.
- I-Mar RT-200 INS for real-time attitude data collection, including heading angle and yaw rate of the vehicle. The sensor has a function to output GPS/INS measurements, and these were post-processed by forward and backward processing using Inertial Explorer for a high accuracy reference.

For the collected irregular driving styles, the post-processed GPS/INS measurements were directly fed into the irregular driving detection algorithm as the reference values. The output of measurements from the Leica Viva GNSS GS15 was used to feed the PFCA/PFCTRA model for positioning and dynamic parameter estimation and then the estimated results were used for the irregular driving detection algorithm.

Besides the collection of irregular driving trajectory data, another important issue is to get the lane's central line data in the field test. In this thesis, the coordinate information of the lane's central line was collected after the initial experiment by a vehicle equipped with i-Mar RT-200 INS (at a 20 Hz output rate) driving along the middle of the lanes on which the irregular driving was conducted. The post-processed GPS/INS measurements for the central line were recognized as the lane's central line coordinate information. The lateral displacement of the vehicle was calculated by the same method as in the simulation, which is searching for the nearest two points of the central line data to the vehicle's location and then calculating the perpendicular distance to the straight line containing these two points. The calculated perpendicular distance is, therefore, the lateral displacement of the vehicle.

In the following sub-sections, the irregular driving detection results from reference and PFCA/PFCTRA estimations are discussed.

5.4.2 Reference Data for Real Irregular Driving Detection

The post-processed irregular driving trajectory measurements for the routes are considered as the reference data. The irregular driving detection algorithm is applied on the reference data of two routes separately with 1, 5 and 10 Hz output rates.

The example of the calculated O-indicator and D-indicator and the corresponding risk type indicator values from the reference data for the first 5 s for session 1 is shown in Figs. 5.56 and 5.57. Tables 5.38 present the example of the first 5 s of weaving driving style detected on the straight lane with the driving classification indicator complying with the rule AB > BC > AD > CD.

The example of the first 5 s of O-indicator, D-indicator and risk type indicator values for session 2 is shown in Figs. 5.58 and 5.59. Tables 5.38 implies the rule CD > BC > AD > AB for session 2, which is detected as jerky driving.

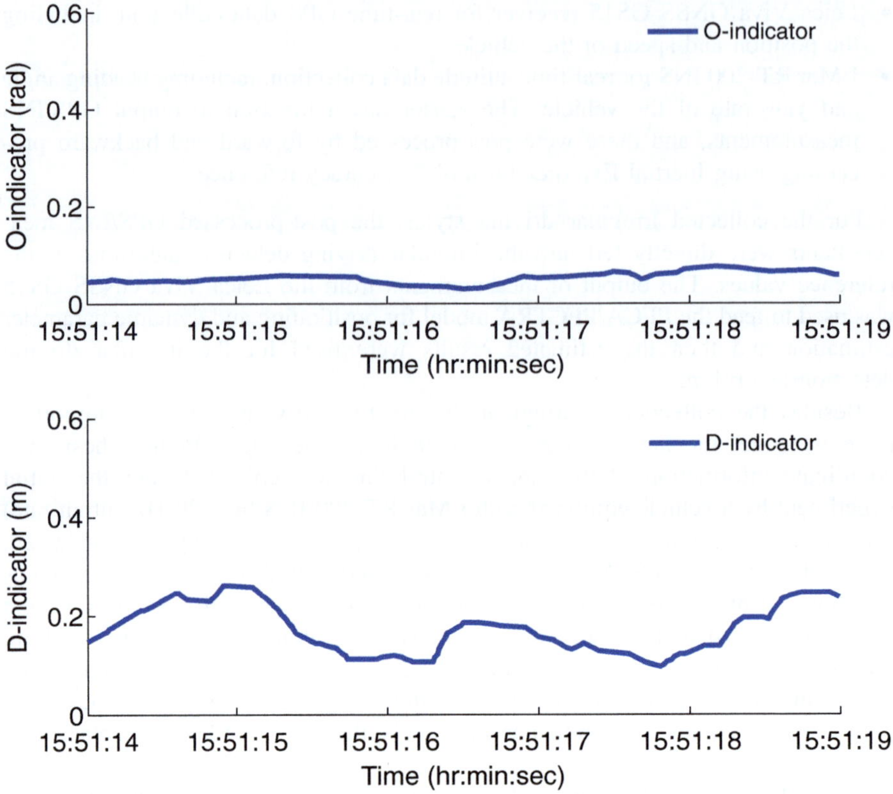

Fig. 5.56 Example of O-indicator and D-indicator from the reference data for session 1 (first 5 s)

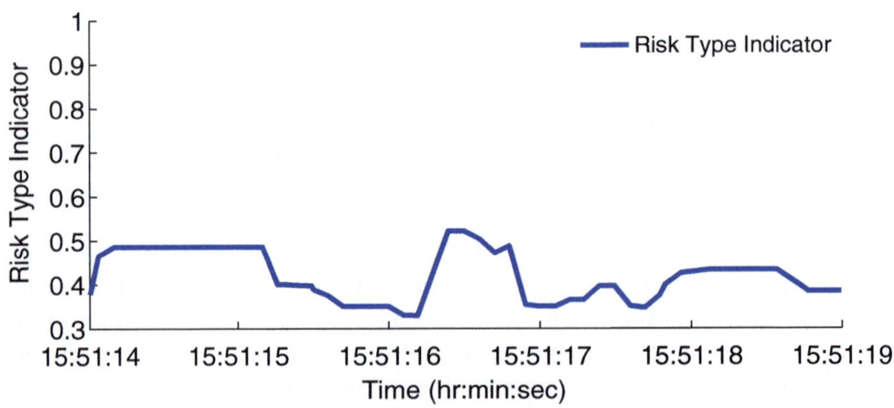

Fig. 5.57 Example of risk type indicator from the reference data for session 1 (first 5 s)

Table 5.38 Driving classification indicators for the first 5 s of five sessions from reference data

Sessions	Five seconds of sessions	AB	BC	CD	AD	Judgement
1	15:51:14.0–15:51:18.9	47	44	3	6	Weaving on straight
2	15:51:41.5–15:51:46.4	0	31	50	19	Jerky driving on curve
3	15:57:13.9–15:57:18.8	0	9	50	41	Jerky driving on straight
4	16:12:10.9–16:12:15.8	29	31	21	19	Swerving on curve
5	16:12:36.4–16:12:41.3	50	41	0	9	Weaving on curve

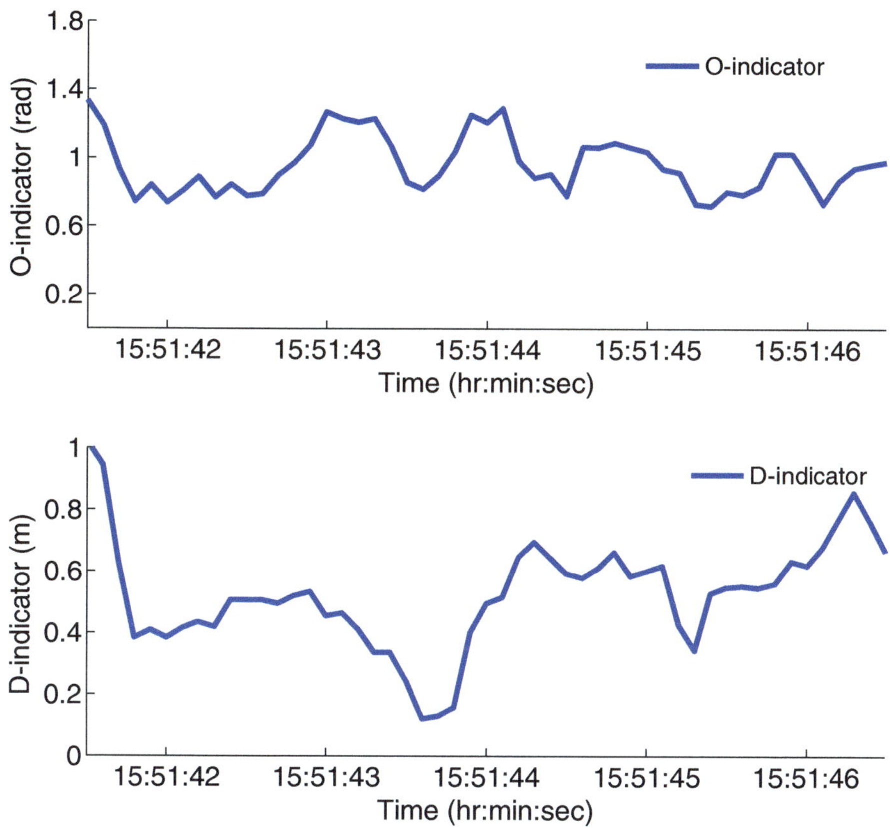

Fig. 5.58 Example of O-indicator and D-indicator from the reference data for session 2 (first 5 s)

Figures 5.60 and 5.61 present the 5 s example of O-indicator and D-indicator results and risk type indicator values for session 3. In Table 5.38, the corresponding first 5 s driving classification indicator for session 3 follows CD > AD > BC > AB, which is detected as jerky driving on a straight lane.

An example of 5 s O-indicator and D-indicator results, and risk type indicator values, for session 4 is shown in Figs. 5.62 and 5.63. Table 5.38 presents the

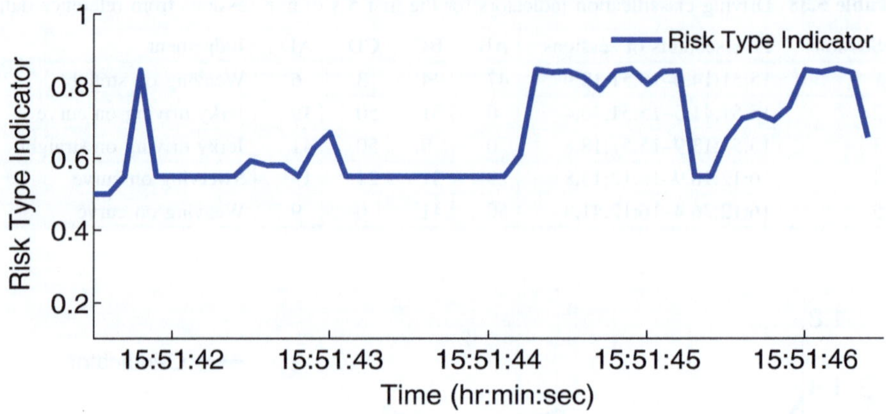

Fig. 5.59 Example of risk type indicator from the reference data for session 2 (first 5 s)

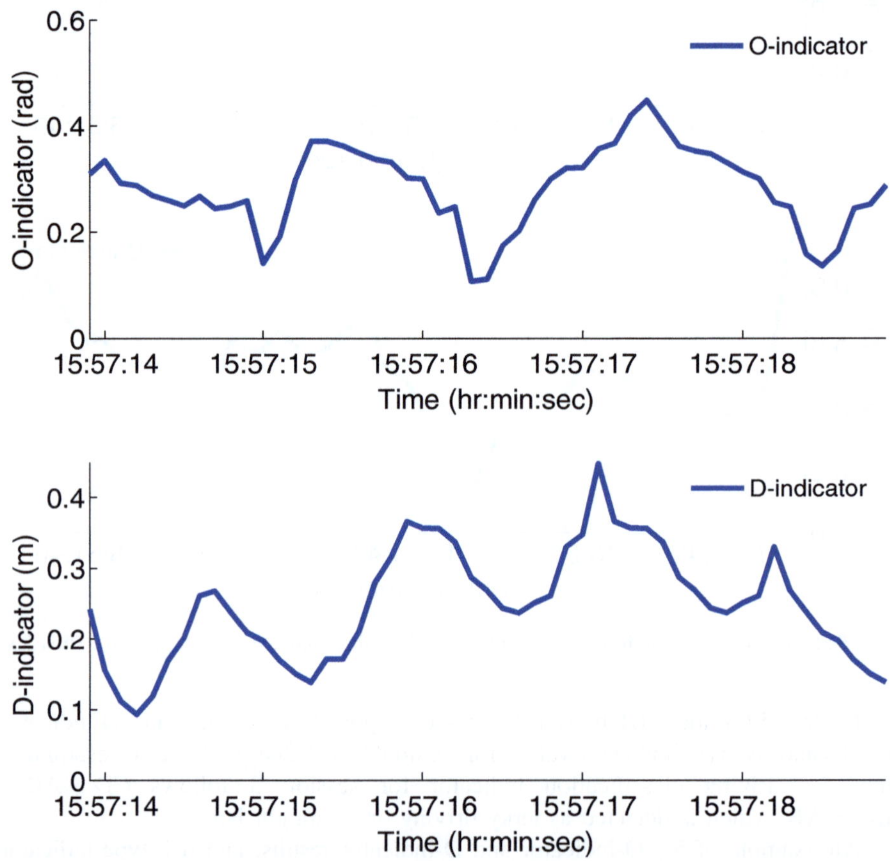

Fig. 5.60 Example of O-indicator and D-indicator from the reference data for session 3 (first 5 s)

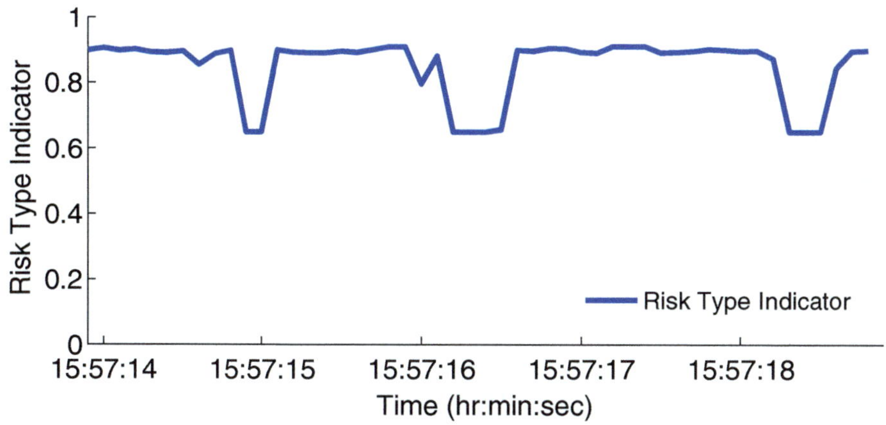

Fig. 5.61 Example of risk type indicator from the reference data for session 3 (first 5 s)

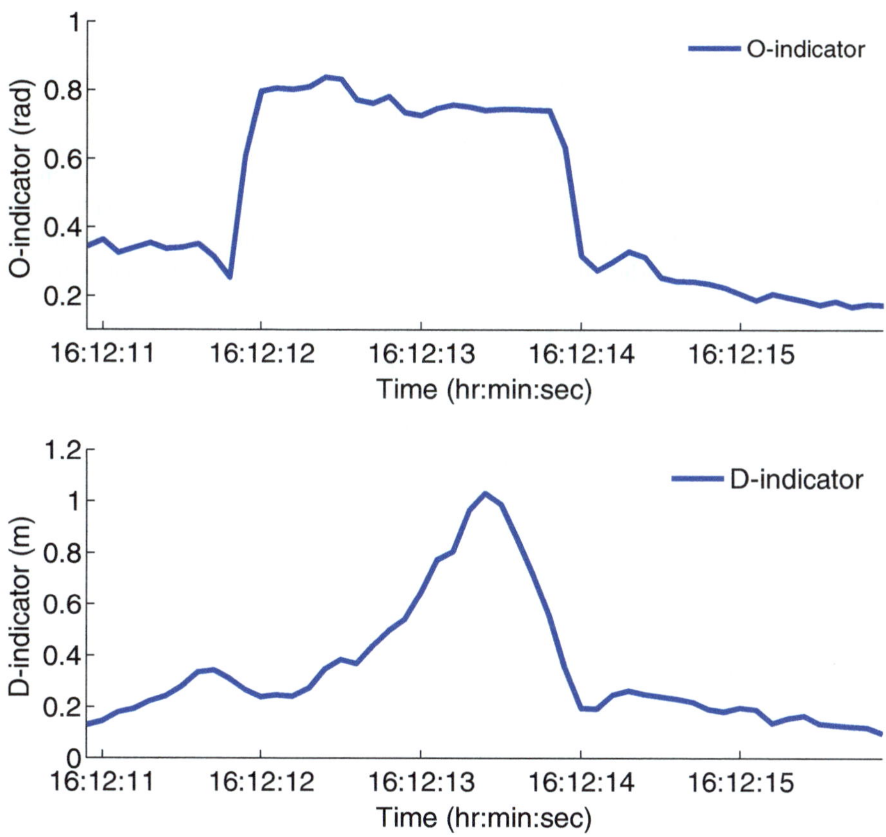

Fig. 5.62 Example of O-indicator and D-indicator from the reference data of session 4 (first 5 s)

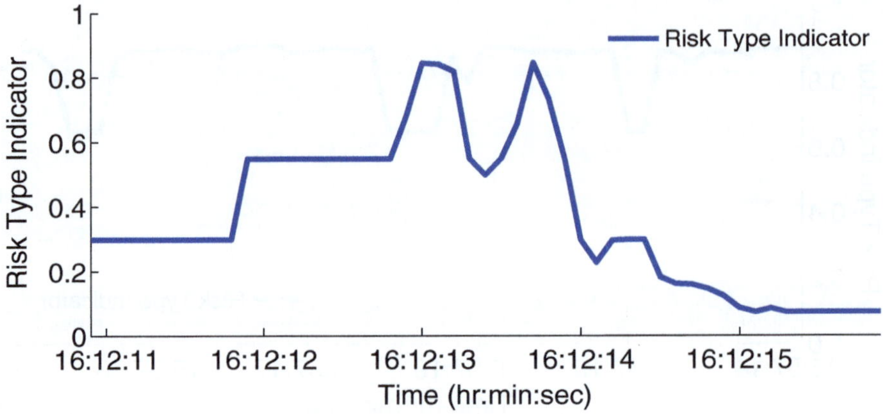

Fig. 5.63 Example of risk type indicator from the reference data of session 4 (first 5 s)

driving classification indicator values complying with the rule BC > AB > CD > AD, which implies swerving on a curve.

An example of O-indicator and D-indicator values and risk type indicator for the first 5 s of session 5 is shown in Figs. 5.64 and 5.65. In Table 5.38, the first 5 s of session 5 is detected as AB > BC > AD > CD, which is weaving on a curve.

The time to first detection for irregular driving based on the various output rates is shown in Table 5.39. It is clear that all of the sessions can be detected within 5 s of the irregular driving happening. The reference data based field test results show that irregular driving can be detected on the routes. It also indicates that the higher the output rate, the earlier the detection time epoch, which is identical with the simulation results. Table 5.40 shows the availability and correct detection rate based on reference results with respect to different output rates. It indicates that the availability and correct detection rate for the reference data increases as the output rate decreases.

From the post-processed GPS/INS combined reference data based irregular driving detection results it is clear that the defined irregular driving styles can be distinguished on a motorway. In the Sect. 5.4.3, the RTK GPS measurements with the PFCA and PFCTRA models that were used for the detection of irregular driving detection are discussed.

5.4.3 Filter Estimated Results for Real Irregular Driving Detection

In this section, NovAtel RTKNav software has been used to output the collected Leica GS15 GPS data. The reference station data is from TEDD in Teddington, which is the nearest station for our test. The output contains the positioning and

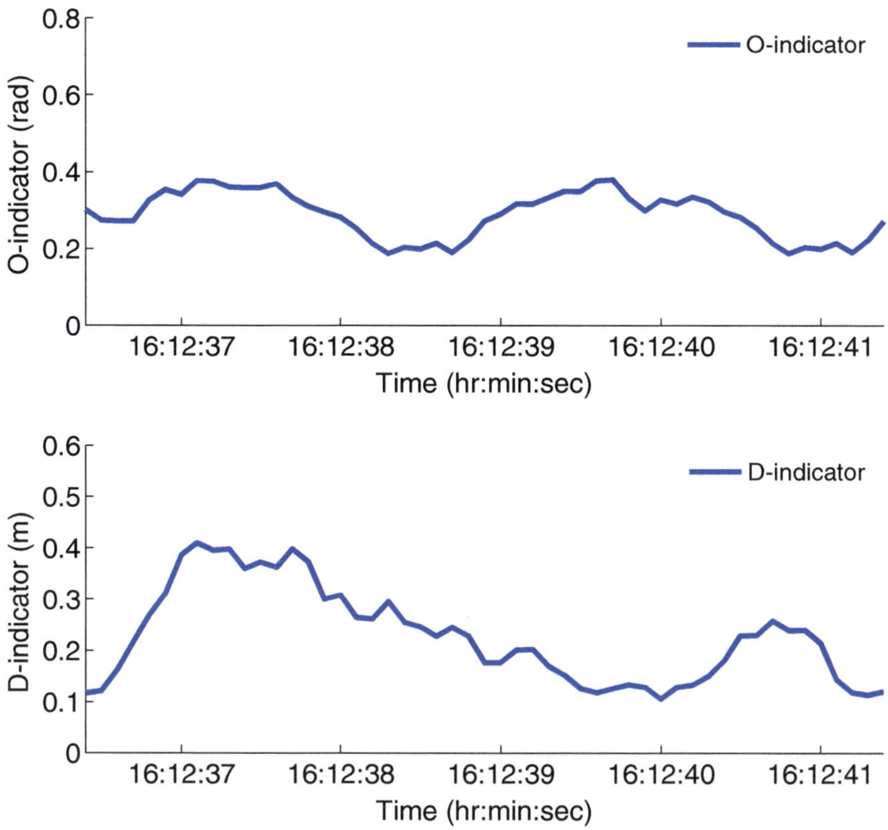

Fig. 5.64 Example of O-indicator and D-indicator from the reference data of session 5 (first 5 s)

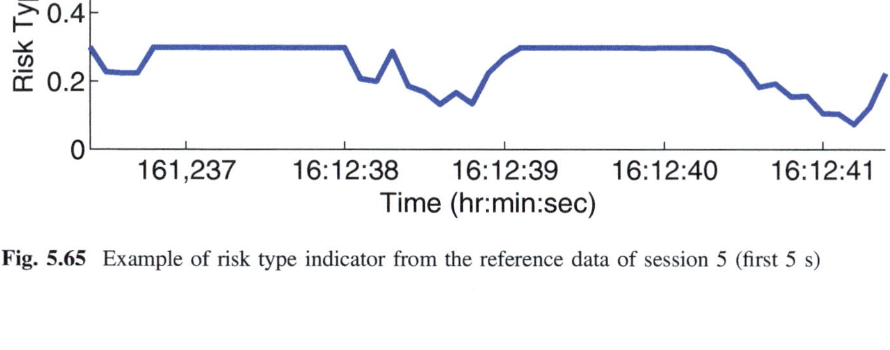

Fig. 5.65 Example of risk type indicator from the reference data of session 5 (first 5 s)

Table 5.39 The time to first detection of irregular driving for sessions based on the output rate from reference results

Output rate (Hz)	Time to first detection				
	Session 1	Session 2	Session 3	Session 4	Session 5
10	15:51:16.2	15:51:45.8	15:57:15.3	16:12:14.5	16:12:39.6
5	15:51:16.5	15:51:46.0	15:57:15.5	16:12:14.5	16:12:40.0
1	15:51:17.0	15:51:46.0	15:57:16.0	16:12:15.0	16:12:40.0

Table 5.40 The availability and correct detection rate based on reference results with respect to different output rate

Output rate (s)	Availability (%)	Correct detection rate (%)
0.1	87.12	87.35
0.5	93.27	94.05
1	95.34	96.75

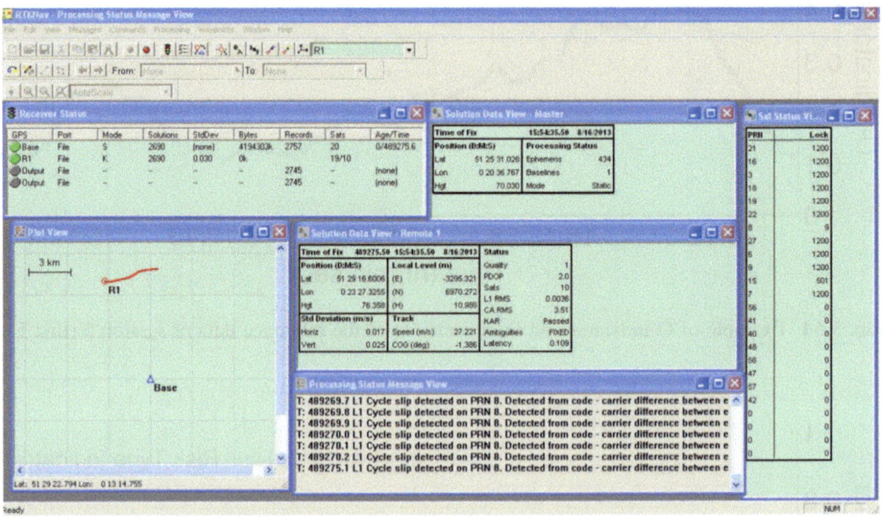

Fig. 5.66 Reply mode for RTKNav real-time output

velocity information of the vehicle. Figure 5.66 shows the reply mode of single baseline RTK processing for the real-time output of the collected GPS data in the RTKNav software.

The RTKNav output different types of positioning solutions, which are shown in Table 5.41. From Table 5.41, it is shown that 84.04 % of the total points were determined with the RTK fixed ambiguity solution in route 1, which is claimed to

Table 5.41 RTKNav positioning solutions statistics (the numbers defined by the National Marine Electronics Association are as follows: 0 for no solution; 1 for single point positioning; 2 for DGPS positioning; 3 for PPS DGPS positioning; 4 for RTK ambiguity fixed positioning and 5 for RTK float ambiguity solution)

Name	Solution type					
	0	1	2	4	5	Total
Route 1	100	0	525	3529	45	4199
Percentage	2.38	0	12.50	84.04	1.08	100
Route 2	52	0	448	927	372	1799
Percentage	2.89	0	24.90	51.53	20.68	100

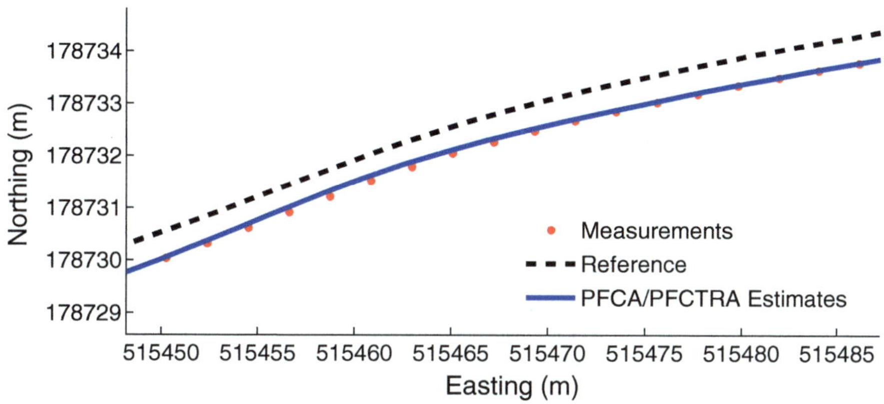

Fig. 5.67 Comparison of RTKNav output GPS positioning and PFCA/PFCTRA estimated positioning results with respect to reference

have sub-metre mean positioning accuracy. The mean positioning accuracy for route 1 was 0.9664 m. In route 2, however, only 51.53 % of the points were determined with the RTK fixed ambiguity solution and the mean positioning accuracy for route 2 was 1.0090 m. As is discussed in the simulation, a 0.5 m positioning accuracy is required for the detection of irregular driving and, therefore, the current RTKNav output positioning is insufficiently accurate. For this reason, the PFCA and PFCTRA models, which showed the smallest error in the straight and curved positioning estimation in the simulation, are used for the real sessions in order to improve the RTK GPS positioning accuracy.

From Fig. 5.67, it is clear that the filter estimated results are closer to the reference position solutions, demonstrating the benefit of filtering in improving the accuracy of the RTK GPS results. Table 5.42 shows the positioning accuracy

Table 5.42 Statistics of the positioning accuracy

Name	Time	Positioning type	Accu. 0–0.5 m	Accu. 0.5–1 m	Accu. 1–1.5 m	Accu. Above 1.5 m	Total
Route 1	15:51:00–15:57:59.9	RTKNav output no. of points	48	1498	2633	20	4199
		PFCA/PFCTRA estimated no. of points	305	3455	422	17	4199
		RTKNav output percentage	1.14	35.68	62.70	0.48	100
		RTKNav output mean accuracy	0.9664 m				
		PFCA/PFCTRA estimated percentage	7.26	82.28	10.06	0.40	100
		PFCA/PFCTRA estimated mean accuracy	0.6214 m				
Route 2	16:10:00–16:12:59.9	RTKNav output no. of points	96	481	1179	43	1799
		PFCA/PFCTRA estimated no. of points	375	1116	277	31	1799
		RTKNav output percentage	5.33	26.72	65.56	2.39	100
		RTKNav output mean accuracy	1.0090 m				
		PFCA/PFCTRA estimated percentage	20.84	62.03	15.41	1.72	100
		PFCA/PFCTRA estimated mean accuracy	0.7883 m				

performance statistics for route 1 and route 2. It is obvious that the PFCA/PFCTRA model estimated positioning solutions have significantly improved the positioning accuracy compared to the RTKNav output results, with the mean positioning accuracy in route 1 being reduced to 0.6214 from 0.9664 m and the mean positioning accuracy in route 2 being reduced to 0.7883 from 1.009 m. The percentage of measurements with a positioning accuracy within 0–0.5 m in route 1 increased from 1.14 to 7.26 % based on the PFCA/EKFCTRA estimations, while the corresponding percentage for route 2 increased from 5.33 to 20.84 %. Furthermore,

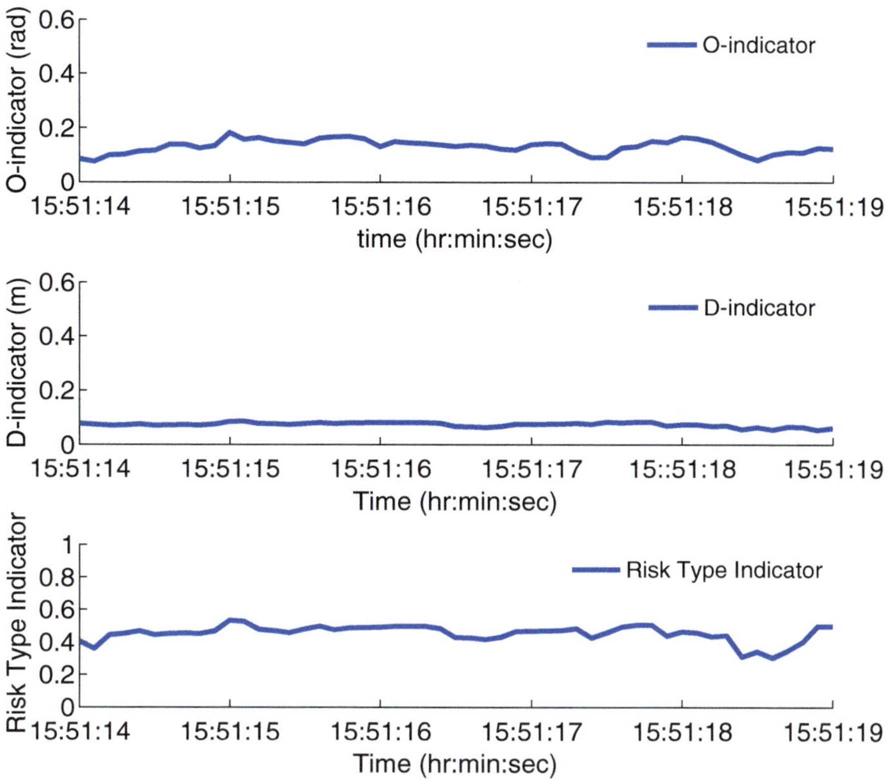

Fig. 5.68 The example of O-indicator, D-indicator and risk type indicator from the PFCA/PFCTRA estimated results for session 1 (first 5 s)

from the analysis of positioning accuracy against the time epochs of the defined sessions, it is shown that the mean positioning accuracy is below 0.5 m after the PFCA/PFCTRA model was applied.

Based on the PFCA/PFCTRA estimated positioning results, the examples of O-indicator and D-indicator values and corresponding risk type indicator results for the first 5 s of the five sessions, are plotted in Figs. 5.68, 5.69, 5.70, 5.71 and 5.72. Table 5.43 presents the 5 s example of driving classification indicators and the judgement from the PFCA/PFCTRA estimated results. From the results of the first 5 s of the five sessions, the detection of different types of irregular driving results are identical to the reference data based irregular driving detection results.

The comparison of the time to the first detection of irregular driving for both reference and PFCA/PFCTRA estimated results with respect to different output

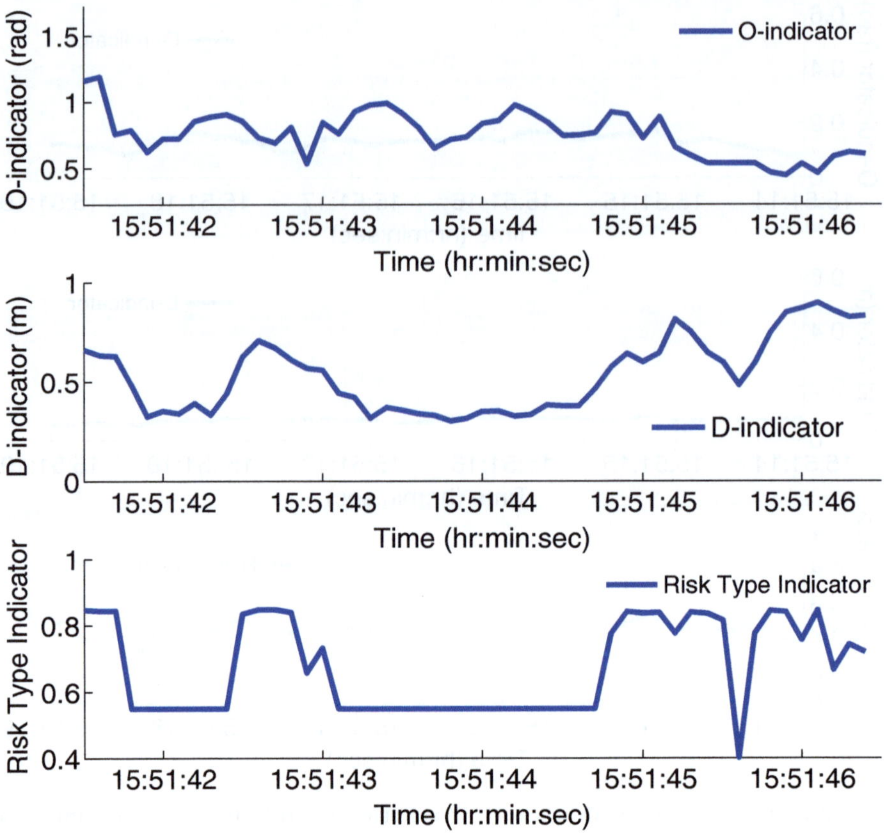

Fig. 5.69 The example of O-indicator, D-indicator and risk type indicator from the PFCA/PFCTRA estimated results for session 2 (first 5 s)

rates for session 1–5 is shown in Table 5.44. In general, the time to first detection from PFCA/PFCTRA is similar to that from the reference. It is also shown, however, that the time to first detection from the PFCA/PFCTRA estimated results can be later than it is from the reference, especially at higher output rates. For example, in Session 4, the time to first detection from the PFCA/PFCTRA estimated results is 0.3 s later than it is from reference with a 10 Hz output rate.

The comparison of availability and correct detection rate of the irregular driving detection algorithm from PFCA/PFCTRA model estimated results and from the reference with respect to the different output rates is shown in Table 5.45. Generally, the availability and correct detection rate decreases as the output rate

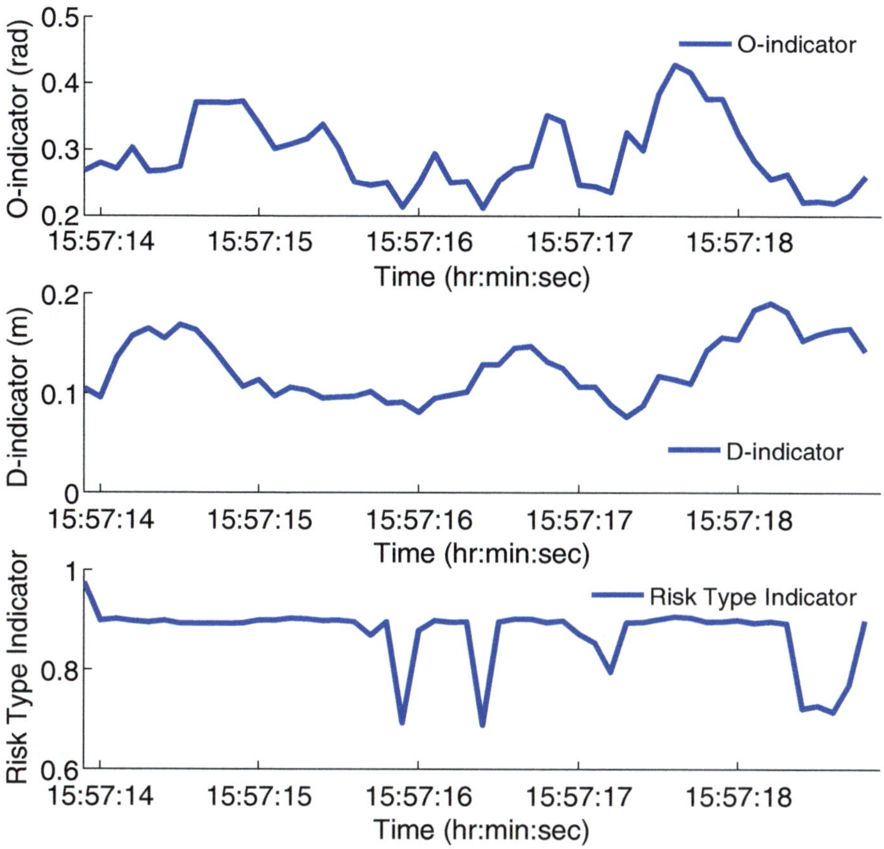

Fig. 5.70 The example of O-indicator, D-indicator and risk type indicator from the PFCA/PFCTRA estimated results for session 3 (first 5 s)

increases. The availability based on the PFCA/PFCTRA estimated results is slightly lower than the availability from the reference results. The correct detection rate from the PFCA/PFCTRA estimations is significantly lower than from the reference, since the PFCA/PFCTRA estimations have more errors than the reference.

From the analysis of the sessions above, the 0.5 m positioning accuracy requirement for the irregular driving detection algorithm is validated in the field test. There are a number of open issues associated with real testing, however. Low quality measurements are always collected when the signal is starting to be recaptured after a loss of signal when the vehicle has just passed under a bridge on

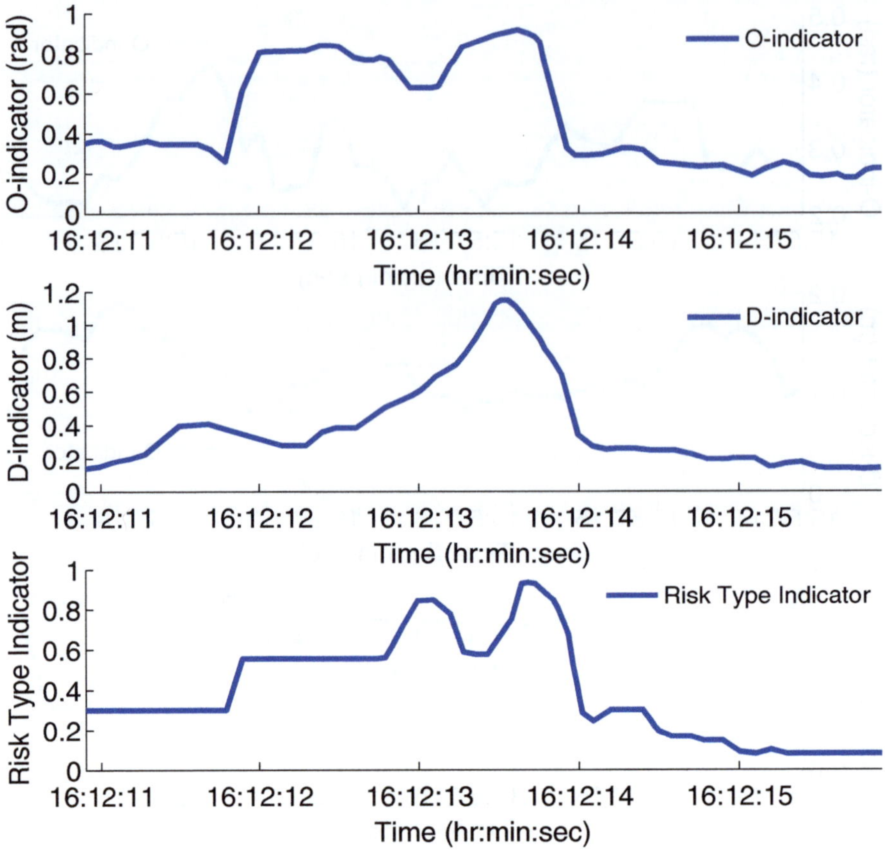

Fig. 5.71 The example of O-indicator, D-indicator and risk type indicator from the PFCA/PFCTRA estimated results for session 4 (first 5 s)

the motorway or is passing by proximate objects such as large lorries and trees. Although the PFCA/PFCTRA estimated GPS positioning results improve positioning accuracy, due to the low quality measurement data, it is still impossible to control the positioning accuracy below 0.5 m for some measurements (as is required for the irregular driving detection algorithm). This is also the reason why the PFCA/PFCTRA estimations have a lower correct detection rate and availability than the reference data, especially in the high rate output.

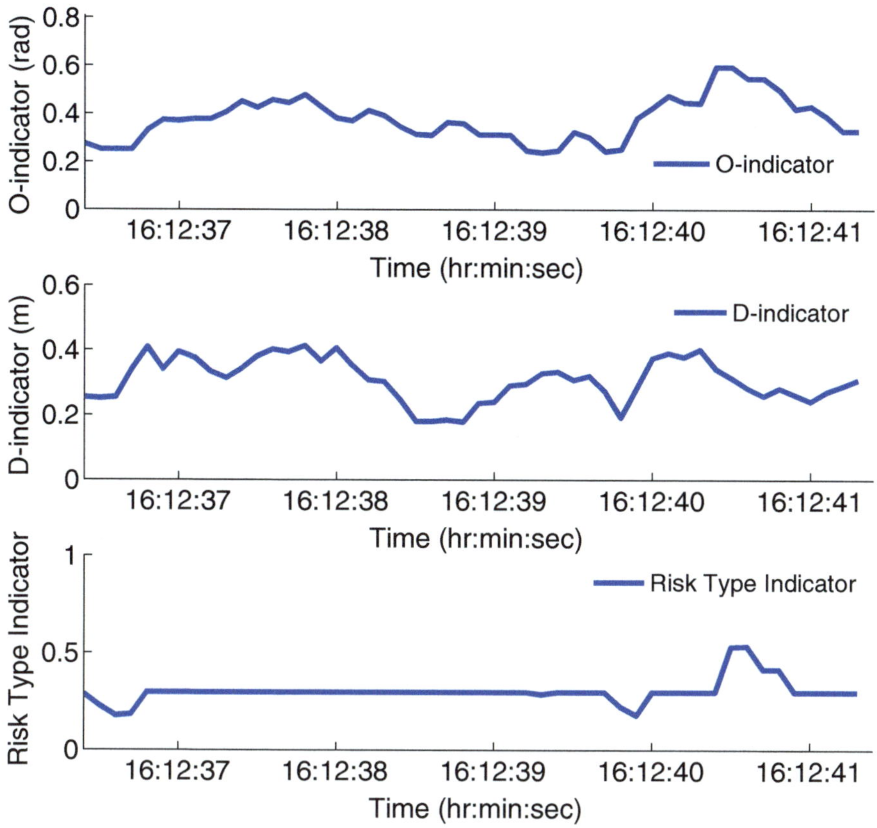

Fig. 5.72 The example of O-indicator, D-indicator and risk type indicator from the PFCA/PFCTRA estimated results for session 5 (first 5 s)

Table 5.43 The example of driving classification indicator for the first 5 s of five sessions from PFCA/PFCTRA estimated results

Session	AB	BC	CD	AD	Judgement
1	49	46	1	4	Weaving on straight
2	0	27	49	23	Jerky driving on curve
3	0	5	50	45	Jerky driving on straight
4	27	31	25	20	Swerving on curve
5	48	44	2	6	Weaving on curve

Table 5.44 The comparison of the time to first detection from the PFCA/PFCTRA estimated results and reference results with respect to different output rate

Output rate (Hz)	Session 1	Session 2	Session 3	Session 4	Session 5
Time to first detection (PFCA/PFCTRA estimated)					
10	15:51:16.4	15:51:45.9	15:57:15.3	16:12:14.8	16:12:39.6
5	15:51:16.5	15:51:46.0	15:57:15.5	16:12:15.0	16:12:40.0
1	15:51:17.0	15:51:46.0	15:57:16.0	16:12:15.0	16:12:40.0
Time to first detection (reference)					
10	15:51:16.2	15:51:45.8	15:57:15.3	16:12:14.5	16:12:39.6
5	15:51:16.5	15:51:46.0	15:57:15.5	16:12:14.5	16:12:40.0
1	15:51:17.0	15:51:46.0	15:57:16.0	16:12:15.0	16:12:40.0

Table 5.45 The comparison of availability and correct detection rate from the PFCA/PFCTRA estimated results and reference results with respect to different output rates

Output rate (s)	Availability (PFCA/PFCTRA estimated) (%)	Availability (reference) (%)	Correct detection rate (PFCA/PFCTRA estimated) (%)	Correct detection rate (reference) (%)
0.1	86.13	87.12	79.82	87.35
0.5	92.35	93.27	89.20	94.05
1	94.21	95.34	94.71	96.75

5.5 Summary

The lane level irregular driving detection algorithm developed in this thesis is evaluated in this Chapter based on simulation and real data.

In the simulation, two types of data are generated based on the defined scenarios (weaving/weaving on curve, swerving/over/under turning, jerky driving/jerky driving on curve and normal/normal on curve). The first type of data is the reference data, used for the definition of scenarios and extraction of the sorting rules from the driving classification indicator for further irregular driving detection. From the sensitivity analysis of the irregular driving detection algorithm, based on the reference data, it can be observed that a higher output rate of the irregular driving detection could lead to an earlier time to first detection for a specific driving style. In addition, the availability and correct detection rate decreases with an increase in the output rate of the system.

The second type of data is the simulated GPS and INS measurements data, which is used to test the performance of the designed lane level precise positioning algorithm and the irregular driving detection algorithm. The positioning results show that the PFCA model results in the highest estimation accuracy in straight scenarios and the PFCTRA model the highest accuracy in curved scenarios. It is also shown that the positioning accuracy performance of the PFCA and PFCTRA models is below 0.5 m for the designed driving styles in straight and curved scenarios respectively. Also, it is shown that the PFCA and PFCTRA estimated results can output the same irregular driving detection results as from the reference data. It is, therefore, confirmed that the PFCA and PFCTRA based estimations can provide accurate positioning and parameter estimations to detect different types of irregular driving. Thus, it is concluded that 0.5 m (95 %) positioning accuracy is the requirement for the irregular driving detection algorithm from the simulation.

The field test has validated that the designed FIS based irregular driving detection algorithm can detect different types of irregular driving on the motorway. It is also validated that filtering estimated results through PFCA/PFCTRA can improve the accuracy of collected GPS data and detect the defined irregular driving sessions. A 0.5 m mean positioning accuracy is also confirmed as the requirement for the irregular driving detection algorithm. The field test results also validate that a higher output rate could lead to an earlier time to first detection and a decrease in the availability and correct detection rate. The results of the field test show that, based on the PFCA/PFCTRA estimated irregular driving detection system, the time to first detection can be within 5 s and the availability and the correct detection rate can reach 94.21 % and 94.71 % respectively with a 1 Hz output rate.

References

Clanton, J., Bevly, D., & Hodel, A. (2009). A low-cost solution for an integrated multisensory lane departure warning system. *IEEE Transactions on Intelligent Transportation Systems, 10*(1), 47–59.

PTV VISSIM. (2013). http://vision-traffic.ptvgroup.com/en-us/products/ptv-vissim/

Chapter 6
Conclusion and Recommendations for Future Work

Abstract This chapter starts by recalling these objectives, and then summarises the main results before drawing the relevant conclusions. Based on the findings of this thesis, recommendations are made for future research in the field of identifying irregular driving based on driving pattern recognition.

The primary aim of this thesis was to develop a system to improve the accuracy of the estimation of vehicle positioning and dynamic parameters by integrating precise vehicle models with GPS/INS based positioning measurements and then to use this in tandem with an designed irregular driving detection algorithm to recognise different types of irregular driving patterns.

A number of specific objectives were formulated to achieve this aim. This chapter starts by recalling these objectives, and then summarises the main results before drawing the relevant conclusions. Based on the findings of this thesis, recommendations are made for future research in the field of identifying irregular driving based on driving pattern recognition.

6.1 Research Objectives

Chapter 1 presented the broad context and aim of this research and formulated seven research objectives. These are re-stated here to facilitate their mapping to the conclusions drawn.

- Carry out an extensive literature review to capture the state-of-the-art in the existing irregular driving monitoring algorithms based on the detection and recognition of vehicles' real-time driving patterns and monitoring of driver behaviours.
- Identify and provide solutions to address basic underpinning issues such as system design, sensor integration, vehicle motion model choice and different driving pattern detection methods.

R. Sun, *An Integrated Solution Based Irregular Driving Detection*,
Springer Theses, DOI 10.1007/978-3-319-44926-5_6

- Develop advanced EKF and PF models for GPS and INS integration with different motion models for precise vehicle positioning and dynamic parameter estimation.
- Develop an advanced FIS algorithm for the classification of driving patterns.
- Establish a Matlab and Simulink based simulation platform with an efficient data processing structure for testing irregular driving identification systems based on multi-combination models.
- Provide analysis of the performance of the novel irregular driving identification algorithms based on real tests, employing both straight lane and curved lane data, and simulated/or real data where practical.

6.2 Conclusions

This thesis has developed an integrated solution based irregular driving detection algorithm. The conclusions are divided into five parts and are presented below.

6.2.1 Road Safety and Intelligent Transport Systems

A review of the literature has shown that the increasing number of vehicles on the roads poses a growing risk to safety. Irregular driving is one of the most significant causes of road traffic accidents, but traditional safety measures are limited in their ability to address this problem. New ITS based methods have the potential to make a positive impact on the problems caused by irregular driving. ITS architecture and applications are reviewed with the conclusion that the detection of irregular driving requires access to high accuracy positioning and vehicle dynamic parameters, and algorithms for the extraction of erratic driving styles.

6.2.2 State-of-the-Art in Irregular Driving Detection

Irregular driving styles, including weaving, swerving, and jerky driving, can cause traffic accidents. Detecting irregular driving at an early stage and issuing timely warnings are therefore essential to avoid accidents.

A significant amount of research has been dedicated to the development of irregular driving detection systems to improve driving safety. Some of this research focuses on the detection and recognition of a vehicle's real-time driving patterns. Other research focuses on monitoring the driver's physical behaviour to prevent irregular driving caused by fatigue or inattention. Analysis of the current research has determined that it is not feasible to detect irregular driving based purely on the physical monitoring of drivers; however, it is feasible to detect irregular driving by detection of the vehicle's real-time driving patterns. Currently, most of the irregular driving detection systems are still in the early stages of development and have not quantified the system performance. In addition, the literature review has shown that none of the existing algorithms by themselves are appropriate for identifying the different types of irregular driving. The reason is that detection performance is currently limited by the availability of high accuracy positioning. Also, some of the current research is just focused on improvements in the GNSS positioning technology while other research is focused narrowly on driving pattern recognition with only rare links between the two.

In order to design the lane level precise positioning algorithm and an irregular driving detection algorithm, the current in-car positioning technology, which has the potential to provide high accuracy positioning for vehicle, has been reviewed. This technology involves vehicle motion sensors (such as GPS and INS), vehicle motion models (such as CV, CA, CTRV and CTRA), and sensor integration models (such as EKF and PF). The three main driving pattern recognition methods are then discussed. The FIS method, which does not need large samples for operation, was judged to be especially suitable for the detection of irregular driving.

6.2.3 A New System for Lane Level Irregular Driving Detection

In order to address the shortcomings discussed in Sect. 6.2.2, this thesis has developed a new algorithm which links highly precise positioning estimation with irregular driving detection. The newly designed estimation model for positioning and vehicle dynamic parameters is realized by exploiting precise motion models and PF/EKF filters for GPS/INS integration and a new irregular driving detection algorithm with a design based on FIS.

The innovations of the designed system are illustrated as follows. The first innovation is that the proposed method takes advantage of merging high accuracy vehicle positioning and dynamic parameter estimations with different types of irregular driving identification technology, a combination which has not been considered by previous investigations. The second innovation is that the new EKF/PF algorithm is designed for GPS/INS integration with motion models for the estimation of precise positioning and *omega* and *d* parameters. The third innovation is that a new FIS based algorithm is developed to classify different types of irregular driving styles.

The framework of the new system can be divided into two parts. The first part is for precise positioning and parameter estimation. In order to collect the vehicle's driving data, one INS with one gyro and one accelerometer mounted along the vehicle body axis is used to output the yaw rate, acceleration and heading for the vehicle's heading direction. One GPS is used to collect the vehicle's local coordinates and heading velocity. The collected initial position is used to feed the PF/EKF models with the precise vehicle motion models to provide the prediction of the positioning and attitude parameters for the next time epoch for iteration. The second part is irregular driving detection. This starts from the calculation of an O-indicator and D-indicator based on the estimated *omega* and *d* of the first part. These two indicators are then used as the input for the FIS based driving pattern identification. The FIS outputs the risk type indicator. In order to amplify the features of each driving style, a driving classification indicator is then developed based on the risk type indicator. Finally, by comparing the sorting of the calculated driving classification indicators with predefined sorting rules extracted from the reference data, the system outputs the identified driving style types.

As described in the system framework, the details of the vehicle kinematic motion and filter models are further designed for the estimation of positioning and related parameters (*omega* and *d*). These designed models include EKFCV, EKFCA, PFCV and PFCA for the straight scenarios and EKFCTRV, EKFCTRA, PFCTRV and PFCTRA models for the curved scenarios. The details of the irregular driving detection algorithm include the design of the O-indicator, D-indicator and risk type indicator membership functions in FIS and also the corresponding mapping rules for the input and output in FIS.

6.2.4 Test, Analysis and Performance Validation

The newly developed algorithms in Sect. 6.2.3 are further tested in simulated irregular driving scenarios and real field sessions.

The simulated irregular driving scenarios include weaving/weaving on curve, swerving/swerving on curve, jerky driving/jerky driving on curve and normal driving/normal driving on curve. In the simulation, there are two types of data generated. The first type of data is reference data, which is for the definition of scenarios and the extraction of the sorting rules from the driving classification indicator for further irregular driving detection. The second type of data is the simulated GPS and INS measurement data, which is used to test the designed precise positioning algorithms and to calculate the sorting of the driving classification indicator to compare with the reference and find out the detection results. A performance assessment of the simulation will be discussed in what follows.

A sensitivity analysis of the irregular driving detection algorithm based on the reference data reveals that the algorithm can be used to distinguish the designed episode with the different irregular driving styles in the test. It also indicates that the higher the output rate of the irregular driving detection, the earlier the initial detection time of the specific driving style is likely to be. This is demonstrated by the fact that the algorithm with a 10 Hz output is able to detect a jerky driving style 0.2 s earlier than with a 5 Hz output. One problem shown by the sensitivity analysis is the output availability of the irregular driving algorithm. Unavailable data always occurs at the point where the driving style is changing from one style to another.

Furthermore, as the output rate of the system increases, the availability and correct detection rate will decrease. Thus, choosing the proper output rate for the required situation is critical.

In respect to the positioning accuracy performance of the designed integration model for the simulated GPS and INS measurements, the PFCV and PFCA models exhibit a mean accuracy better than 0.5 m for all the straight scenarios, while PFCTRA exhibits a mean accuracy better than 0.5 m in all curved scenarios. All other models have a mean positioning estimation accuracy below 1 m in all of their scenarios. In addition, the PF model exhibits a smaller error than the EKF model and the CA model exhibits a smaller error than the CV model during the simulation. PFCA and PFCTRA models have the highest positioning accuracy in their scenarios, respectively.

In respect to the performance of the irregular driving detection model in the simulation, the performance is discussed based on the positioning error levels. The EKFCV, EKFCTRV, EKFCA, EKFCRA and PFCTRV models all have a positioning accuracy better than 1 m. The EKFCV and EKFCTRV models, however, wrongly detected swerving/under-turning as jerky driving or weaving, while the EKFCA and EKFCTRA models wrongly detected weaving/weaving on curve as swerving/swerving on curve. Furthermore, EKFCTRA also wrongly detected normal driving as weaving on curve. The PFCTRV model incorrectly detected swerving on curve as weaving on curve. It is obvious that none of the model estimations above can detect the irregular driving scenarios correctly, due to the low level of positioning accuracy. The PFCV, PFCA and PFCTRA models, however, have a mean positioning estimation accuracy better than 0.5 m, and each of these three model based estimations correctly distinguished the different irregular driving scenarios. Thus, from the simulation, it can be concluded that a positioning accuracy of 0.5 m is required for the irregular driving detection algorithm.

A field test was carried out and reveals the following issues.

The field test validated that filtering estimated results through PFCA/PFCTRA can improve the accuracy of collected GPS data and detect the defined irregular driving sessions. A 0.5 m mean positioning accuracy was also confirmed as the requirement for the irregular driving detection algorithm.

The quality of GPS measurements is also important for the performance of the integrated solution based irregular driving detection algorithm. Low quality measurements are always collected when the signal is starting to be recaptured after a loss of signal when the vehicle has just passed under a bridge on the motorway or is passing by proximate objects such as large lorries and trees. Although the PFCA/PFCTRA estimated GPS positioning results improve positioning accuracy, due to the low quality measurement data, it is still impossible to control the positioning accuracy better than 0.5 m for some measurements (as is required for the irregular driving detection algorithm). This is also the reason why the PFCA/PFCTRA estimations have a lower correct detection rate and output availability than the reference data, especially in the high rate output mode.

6.3 Recommendations for Future Research

There are a number of issues that have arisen from this thesis that should be addressed in future research. Some of these issues are unsolved in this research because of a lack of resources or limitations of time; others are out of the scope of this research but are related and would be interesting to investigate in the future.

1. In the field test, only single reference RTK GPS results are fed into the PFCA/PFCTRA algorithm to test the performance of irregular driving in this thesis, with acceptable results for the positioning results. It would be useful to study whether positioning performance could be further enhanced by Network RTK or PPP.
2. The filter algorithms could be tested with multi-satellite systems such as GLONASS/GNSS/BeiDou.
3. Validation schemes for four types of driving styles based on curved and straight roads have been proposed in this thesis. The conclusions as to which scheme is optimal for each application scenario are only based on a specific section of road data from the M4. Further tests should be conducted on different types of roads.
4. The performance of the algorithms should be quantified in representative weather conditions to determine whether practical services can be provided.

From the conclusions and recommendations for further work presented above and the objectives in Sect. 6.1, it is clear that this thesis has achieved its research objectives.

Appendix
Field Test Risk Assessment

Hazardous situation/activity	Potential risk	Control measures
Installation of the GPS and INS equipment on the vehicle	Drop the equipment	Carefully carrying the equipment
		Wearing working boots
Conduct irregular driving manoeuvres on the motorway	Damage of vehicle and equipment when collision with the other vehicles	Look around carefully and ensure no other vehicle around before conducting the irregular driving
	Collision with the infrastructure	Check car condition before doing irregular driving
	Car turn over	Quick stop irregular driving when find other cars coming
	Injury from car accident	Grab the handle of the vehicle when in the car

R. Sun, *An Integrated Solution Based Irregular Driving Detection*,
Springer Theses, DOI 10.1007/978-3-319-44926-5

MIX
Papier aus verantwortungsvollen Quellen
Paper from responsible sources
FSC® C105338

If you have any concerns about our products,
you can contact us on
ProductSafety@springernature.com

In case Publisher is established outside the EU,
the EU authorized representative is:
Springer Nature Customer Service Center GmbH
Europaplatz 3, 69115 Heidelberg, Germany

Printed by Libri Plureos GmbH
in Hamburg, Germany